Viacheslav Karmalita

Metrology of Automated Tests

Also of interest

Digital Processing of Random Oscillations
Viacheslav Karmalita, 2019
ISBN 978-3-11-062500-4, e-ISBN (PDF) 978-3-11-062797-8,
e-ISBN (EPUB) 978-3-11-062517-2

Systems Control Theory
Liu Xiangjie, 2018
ISBN 978-3-11-057494-4, e-ISBN (PDF) 978-3-11-057495-1,
e-ISBN (EPUB) 978-3-11-057527-9

Systems, Automation, and Control
Nabil Derbel, Faouzi Derbel, 2019
ISBN 978-3-11-059024-1, e-ISBN (PDF) 978-3-11-059172-9,
e-ISBN (EPUB) 978-3-11-059031-9

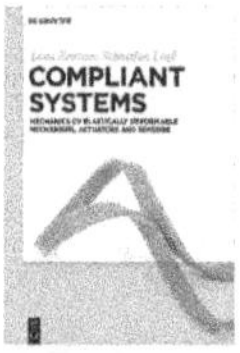

Compliant Systems
Lena Zentner, Sebastian Linß, 2019
ISBN 978-3-11-047731-3, e-ISBN (PDF) 978-3-11-047974-4,
e-ISBN (EPUB) 978-3-11-047740-5

Viacheslav Karmalita

Metrology of Automated Tests

Static and Dynamic Characteristics

DE GRUYTER

Author
Dr. Viacheslav Karmalita
255 Boylan Avenue
Dorval H9S 5J7
Canada
karmalita@videotron.ca

ISBN 978-3-11-066664-9
e-ISBN (PDF) 978-3-11-066667-0
e-ISBN (EPUB) 978-3-11-066669-4

Library of Congress Control Number: 2019958019

Bibliographic information published by the Deutsche Nationalbibliothek
The Deutsche Nationalbibliothek lists this publication in the Deutsche Nationalbibliografie; detailed bibliographic data are available on the Internet at http://dnb.dnb.de.

Cover image: Martin Barraud/OJO Images/Getty Images
Typesetting: Integra Software Services Pvt. Ltd.
Printing and binding: CPI books GmbH, Leck

www.degruyter.com

Preface

When new models of production articles are created, the basic knowledge about these objects comes from testing their prototypes. Such knowledge is obtained by measuring the parameters and characteristics of mechanical, thermophysical, gas dynamic, chemical and others processes that take place in such a prototype.

Progress in the creation of new knowledge as well as in the development of new technical and technological solutions largely depends on the accuracy of measurement results. Therefore, increasing the information content of tests and the reliability of their results are critical issues related to the development, manufacturing and overhaul tests of production objects. An effective solution to these issues, as practice has shown, is provided by the automated measurement-information systems (*AMIS*s) with abilities to control the test object as well as the testing equipment. The *AMIS* allows to:
- receive a large amount of detailed measurement data;
- exclude the influence of human factors on the measurement results;
- ensure the required accuracy of test results.

The metrological characteristics of measurement means (*MM*s) produced by a particular manufacturer are generally well known. In the documentation that accompanies the *MM* there is always information about the measurement errors. This information can be common to the entire production series of the *MM* or it could be specific to a particular sample obtained via calibration on reference equipment. However, even for the *MM* with individual calibration, the measurement error may include components that require verification under specific working conditions.

As a rule, *AMIS*s are created from products of various manufacturers according to technical requirements (*TR*s) including the requirements for measurement accuracy. The adequacy of system's metrological characteristics to the *TR*s is confirmed by a set of metrological studies and procedures for estimation of components of the error arising from individual components of the measurement channel (*MC*) as well as for the entire *MC*. In addition, some object parameters can be determined only through indirect or aggregate measurements, that is, by processing direct measurements of several physical values. In this case, it is necessary to carry out appropriate experimental and computational studies to determine the errors of such measurements. When the test object is in a non-stationary operational mode, it is also necessary to investigate the inertial characteristics of the *MC*s.

The monograph of Dr. Karmalita is devoted to the issues related to the accuracy of measurement results obtained using *AMIS*. It summarizes the author's vast and versatile experience in the field of *AMIS* design and the study of their metrology. After graduating in 1971, Viacheslav Karmalita began his career at the Central Institute of Aviation Motors (*CIAM*) as an experimental engineer for an altitude test facility. During the last decade (1983–1993) of his work at *CIAM*, he headed the

https://doi.org/10.1515/9783110666670-202

department for "Automation of Testing Aircraft Engines and their Components". In addition to scientific research, Dr. Karmalita was engaged in academic activities as a post-graduate lecturer for "Statistical Methods for Processing Experimental Data" course at *CIAM*. Two students under his leadership went on to obtain *Ph.D.* degrees. Viacheslav Karmalita also published several monographs in English and Russian. Currently, he is a well-known expert in the field of *AMIS*s with a specialty in the measurements of stationary and dynamic processes as well as statistical analysis of their results.

Since the measurement results are always probabilistic due to the presence of random components, the first chapter of this book is devoted to the issues of mathematical statistics. It gives a concept of probabilistic models and considers the elements of statistical estimation related to the adaptation of those models to experimental data.

As the given methods and approaches are accompanied by specific examples from the practice of testing gas turbine engines (*GTE*s), the second chapter provides a general idea about the *GTE*, the structures of the *MC* and *AMIS*, and the procedures for acquisition and processing of measurement results.

The third chapter describes the procedures for estimating metrological characteristics of the *MC*s for measuring stationary parameters. An approach to evaluate the accuracy of test results is considered as a logical prolongation of *MC*'s statistical tests and it is shown how to ensure the required accuracy as well.

The last chapter provides information about deterministic and random processes in a dynamic system that is the *MC*. It shows, how through dynamic calibration of the *MC*, the channel inertia may be presented in a formal description. Finally, a method for eliminating the systematic errors of transient process measurements is discussed.

This book is of interest to specialists involved in testing complex manufactured objects and is useful for graduate and post-graduate students with engineering specialties.

October, 2019

Dr. Boris Mineev
Chief Metrologist of CIAM
Moscow, Russia

Contents

Preface — V

Introduction — 1

1 Elements of probability theory and mathematical statistics — 5
1.1 Probabilistic models — **5**
1.1.1 Random variable — **6**
1.1.2 Function of random variables — **12**
1.2 Adaptation of probability models — **18**
1.2.1 Processing of experimental data — **18**
1.2.2 Criterion of maximum likelihood — **21**
1.2.3 Properties of maximum likelihood estimates — **23**
1.2.4 Least-squares method — **27**
1.3 Statistical inferences — **30**

2 Metrological model of automated measurements — 33
2.1 Measurement channel — **35**
2.2 Processing of measurement results — **37**

3 Steady-state performance — 39
3.1 Metrology of measurement channel — **39**
3.1.1 Statistical tests of measurement channel — **40**
3.1.2 Estimation of metrological characteristics — **44**
3.2 Analysis of test result accuracy — **47**
3.2.1 Simulation of measurement errors — **48**
3.2.2 Correlations of test results — **52**
3.3 Ensuring of required accuracy — **59**
3.3.1 Ranking of error sources — **59**
3.3.2 Optimization of accuracy requirements — **63**
3.3.3 Multiple measurements — **67**

4 Transient regimes in test objects — 70
4.1 Processes in dynamic systems — **70**
4.1.1 Deterministic processes — **70**
4.1.2 Random processes — **76**
4.2 Dynamic calibration of measurement channel — **78**
4.2.1 Calibration procedure implementation — **78**
4.2.2 Approximation of frequency response — **84**
4.3 Estimation of dynamic characteristics — **89**

4.3.1 Tikhonov's regularization method — **90**
4.3.2 Recovery of transient processes — **92**
4.3.3 Biases of dynamic characteristics — **96**

References — 103

Index — 105

Introduction

An inalienable part of scientific and engineering activity is cognitive action. It is realized on the basis of formulated cognitive tasks that allow to decompose a research problem into sequential steps. Cognitive tasks may be classified as theoretical and empirical.

An example of a theoretical task is the creation of a mathematical model of the studied object (phenomenon). Empirical tasks consist of disclosure, examination and precise description of facts related to research phenomena. Solutions to empirical tasks are realized by means of such specific cognitive methods as experiment (test) and measurement.

During testing there is a deliberate interference to operational modes of a test object as well as to its operational conditions. Testing allows to find out object's properties (performance) in standard and unconventional operational conditions as well as examining, for instance, the results of changes to the object's design.

Measurement is the cognitive method for obtaining quantified data about research objects (phenomena). It includes two relatively independent procedures: a quantified estimation (measurements) of physical values and an empirical verification of reliability (impartiality) obtained measurement results. The latter issue is a matter of the applied (industrial) metrology ensuring the suitabilty of measurement means via their calibration and quality control.

Tests are an important part of *R&D* stage as well as production of complex manufactured articles. Being a source of knowledge for real processes in the prototypes, the *R&D* tests allow to attain required parametric perfection and reliability of designed products. Experiments are a crucial part of the manufacturing cycle as well – a product's operation is checked and its performance is validated during acceptance tests.

The tests are carried out in dedicated test facilities containing diverse technological (fuel, oil, hydraulic, electrical, communication etc.) systems to support operation of test objects as well as test equipment (stand, object control unit, transducers, actuators etc.). Modern test facilities are characterized by a high level of automation that provide:

- reduction of time and efforts for preparation and implementation of tests;
- mechanization of measurements of test object parameters as well as processing and analysis of experimental data;
- optimization of the test procedure due to its targeted implementation on the basis of real-time information.

Automated tests relieve the test facility staff from time-consuming and routine work for preparation and implementation of tests, thus enabling experimenters to focus on better understanding of test results. Subsequently, the facilitation and acceleration of their work inevitably leads to a qualitatively new organization of tests.

https://doi.org/10.1515/9783110666670-001

The cornerstone of automated tests is the Automated Measurement-Information System (*AMIS*). In addition to increasing the operational efficiency of tests, *AMIS* provides a unified environment for all informational streams and manipulations related to the test procedures (Fig. 1).

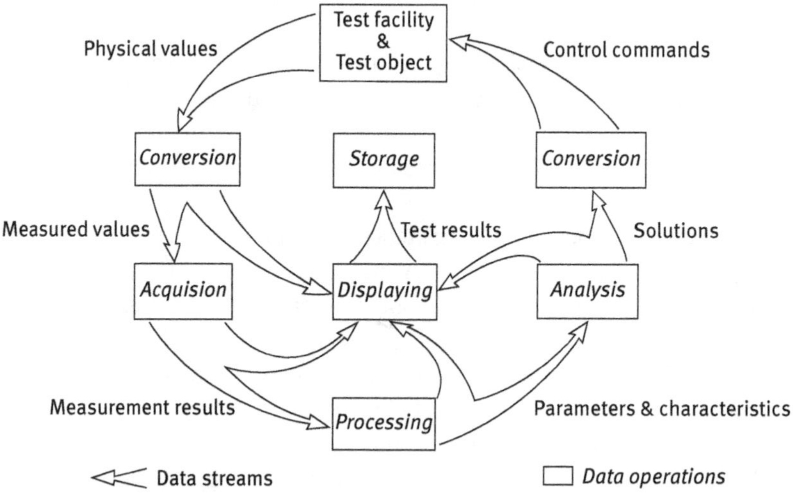

Fig. 1: Information streams/operations related to test procedures.

In this particular case, the term information designates results of measurements and their consequent processing called experimental data in the following text. The data is represented by a set of physical values converted by different sensors/transducers into uniform electrical signals. As a result, information about the status of the test equipment as well as the test conditions and the object itself will be contained in values of electrical currents or voltage, their oscillation frequencies, and so on. The numerical parameter estimation of these measurement signals provides information about the physical values in digital form.

As a rule, the object's parameters and characteristics of power, efficiency, etc. are the result of processing measurement data. Analysis of these parameters/characteristics allows to control the test equipment/object in accordance with test procedure requirements. Necessary adjustments of the test procedure are realized in the forms of dedicated commands for Programmable Logic Controllers (*PLC*s) and equipment/object actuators.

Development, installation and subsequent operation of *AMIS* requires solving numerous tasks linked with different system aspects – instrumentation, hardware, software, operating/processing algorithms, metrology, maintenance support, etc. This book deals with metrological tasks, that is, issues related to the accuracy of test results.

There are several arguments necessitating the discussion of metrological aspects. As it was mentioned earlier, *AMIS* not only creates measurement data (results of measurements) but the system processes these results to evaluate parameters/characteristics of the test objects as well. This principal distinction of *AMIS* from standalone *MMs* requires relevant approaches to estimate accuracy of test results.

As a rule, modern manufactured objects are characterized by high parametric excellence. Hence, attaining the required performance of an object prototype cannot be done through one solution; it is usually realized by a large number of design actions. Each one affects an object's performance on the order of percentage units, sometime less than 1%. This implies that only well-known and ensued accuracy of test results allows to make conclusions about efficiency of undertaken actions in the presence of a statistical variation of experimental data.

In fact, *AMIS* is a subsystem of a complex test system including research and measurement methods as well as different test equipment. Therefore, from the metrological standpoint, the test results contain experimental and measurement errors (Fig. 2).

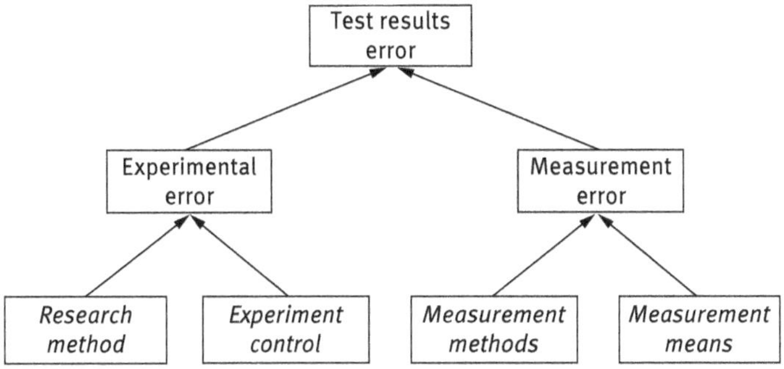

Fig. 2: Components of test result error.

The experimental error is determined by a chosen method of an object's performance estimation and certain grade of control of tests. As a rule, the same characteristic of the test object can be determined with different approaches. Each approach requires a relevant test procedure and object preparation, that is, an appropriate installation of sensors and transducers affecting the object's working processes.

In a well-controlled test, the object's modes and its operational conditions can be augmented exactly as desired, and any influence of external factors is excluded. Poorly controlled tests may lead to attributing effects of unaccounted factors to the object's properties, that is, to distorted data about the real properties of the test object.

The measurement errors contain methodical and instrumental components. The first one is linked with an approach of object's parameter evaluation. For instance, any test object has a 3-dimential structure but parameters of the test object are estimated in a few planes called measurement cross-sections. Further, in each plane, only a finite number of sensors can be installed without significantly affecting the working process of the object. The number of cross-sections and sensors as well as the location of sensors in the planes are sources of methodical errors interfering with the estimates of integral parameters of the object.

The experimental and methodical measurement errors are inseparably linked with the specifics of a test object and its working process as well as the types of measured physical values and applied measurement methods. Due to the diversity of these issues, experimental and methodical errors are not considered in this book.

The instrumental errors are associated only with the utilized measurement means as well as hardware and software of a given *AMIS*. The following text deals with examining the instrumental errors and their influence on accuracy of test results. In particular, this work will describe the approaches and methods to evaluate instrumental errors and their impact on test results for:

- estimating statistical variance caused by instrumental errors;
- eliminating systematic errors (biases) that occur as a consequence of inertial properties of measurement means.

This book material is presented in a form that answers the conceptual questions inherent in a scientific monograph ("Why?") and a handbook ("How?"). It is the latter that explains the variety of examples discussed in the text. These examples confirm the general meaning of concepts of the presented approaches as well as serve to develop skills associated with their implementation.

1 Elements of probability theory and mathematical statistics

As a rule, the measurement of physical values is implemented via conversion of one value to another on a basis of known physical phenomena. For example, the temperature of a substance can be directly converted into electricity through a thermocouple. This conversion uses the Seebeck effect, according to which a thermocouple can generate direct current (*DC*) voltage. To make this work, a junction of two different conductors should be located in the temperature measurement zone, while the other ends should be placed in the reference temperature zone. The magnitude of the *DC* voltage is defined by the Seebeck coefficient depending on the composition of the conductor.

In fact, the implementation of any physical conversion cannot be ideal, since it may be accompanied by various factors which can interfere with conversion results. In the example of a thermocouple, such a factor may be the technological tolerances of conductor production. In other words, manufactured thermocouples of the same type may have different Seebeck coefficients and will therefore generate different voltages at the same temperature. In addition, conductors with current resistance typically have thermal noise that increases with temperature. It associates with the chaotic movement of charge carriers resulting in voltage fluctuations at the ends of the conductor.

In addition to factors inherent to utilized physical phenomena, the surrounding environment can vary drastically since it is never under full control. All mentioned circumstances lead to changes called error in the results of repeated measurements. The error is designated in the subsequent text as Ξ and may have a constant component (bias) as well as variable (random) one.

The bias value is determined during the calibration of the measurement means and can be used as a correction to the measurement results. A random component observed during the calibration procedure must accounted for to allow for its subsequent consideration. It necessitates the discussion of mathematical models related to the concept of probability.

1.1 Probabilistic models

This section deals with models that are mathematical descriptions of random phenomena. The following material is devoted to presentation of such models as well as examination of their properties.

https://doi.org/10.1515/9783110666670-002

1.1.1 Random variable

The error Ξ disperses the results of measurements of a physical value within an area that forms a sample space **A**. Elements (points) of this space may be grouped by different ways into subspaces $A_1, \ldots, A_i, \ldots, A_k$ referred to as events. Appearance of an experimental result inside any subspace implies the occurrence of a specific event. That is to say, the experiment always results in the event:

$$\mathbf{A} = \mathrm{A}_1 + \mathrm{A}_2 + \ldots + \mathrm{A}_k.$$

A certain event A_i may be given a quantitative characteristic through the frequency of this event's occurrence in n experiments. Let $m(\mathrm{A}_i)$ be the number of experiments in which the event A_i was observed. Then the frequency $\nu(\mathrm{A}_i)$ of this event (event frequency) can be determined by the following expression:

$$\nu(\mathrm{A}_i) = m(\mathrm{A}_i)/n.$$

It is evident that the event frequency can be calculated only at the experiment's completion and, generally speaking, depends on the kind of experiments and their number. Therefore, in mathematics an objective measure $P(\mathrm{A}_i)$ of the event frequency is postulated. The measure $P(\mathrm{A}_i)$ is called the probability of the event A_i and is independent on the results in individual experiments. It is possible to state that:

$$P(A_i) = \lim_{n \to \infty} \nu(\mathrm{A}_i).$$

If the experiment result is represented by a real number Ξ called a random variable, one may represent events in a form of conditions $\Xi < \xi$, where ξ is a certain number. In other words, an event may be determined as a multitude of possible outcomes satisfying the non-equality $\Xi < \xi$. The probability of such an event is a function of ξ and is called the cumulative distribution (or just distribution) function $F(\xi)$ of the random variable Ξ:

$$F(\xi) = P(\ \Xi \leq \xi).$$

It is clear that if $a \leq b$, then

$$P(a) \leq P(b);\ P(-\infty) = 0;\ P(+\infty) = 1.$$

Any distribution function is monotonous and non-diminishing. An example of such a function is represented in Fig. 1.1.

If the probability distribution function $F(\xi)$ is continuous and differentiable, its first derivative of the form

$$f(\xi) = \frac{dF(\xi)}{d\xi}$$

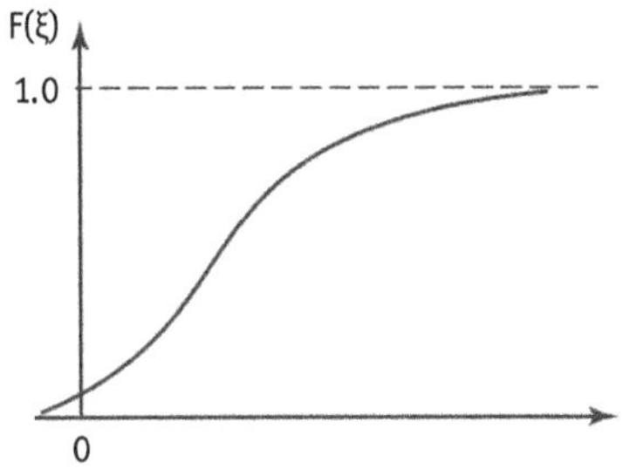

Fig. 1.1: A view of probability distribution function.

is termed as the probability density function (*PDF*) of the random variable Ξ. Note that:

$$P(\Xi \le a) = F(a) = \int_{-\infty}^{a} f(\xi)d\xi;$$

$$P(a \le \Xi \le b) = \int_{a}^{b} f(\xi)d\xi = F(b) - F(a);$$

$$\int_{-\infty}^{\infty} f(\xi)d\xi = 1.$$

In practice these are the parameters of the distribution function that are often used instead of the function itself. One of these parameters is the mathematical expectation of the random variable Ξ:

$$\mu_\xi = M[\Xi] = \int_{-\infty}^{\infty} \xi \cdot f(\xi)d\xi.$$

The expectation of any real, single-valued, continuous function $g(\Xi)$ may be expressed in the similar way:

$$M[g(\Xi)] = \int_{-\infty}^{\infty} g(\xi) \cdot f(\xi)d\xi.$$

Note that mathematical expectations are not random but deterministic values.

Of particular interest are functions of the type:

$$g_l(\Xi) = (\Xi - \mu_\xi)^l,$$

whose expectations are referred to as the lth-order central moments noted as:

$$\alpha_l = M\left[(\Xi - \mu_\xi)^l\right].$$

Specifically, the value $\alpha_2 = D_\xi = \sigma_\xi{}^2$ is the lowest-order moment which evaluates the mean deviation of a random variable around its expectation. This central moment is called variance and σ_ξ is referred to as the root-mean-square (*rms*) deviation.

As an example, let us examine the probability density of random variables referred to in this text. The first example is related to a probability scheme characterized by maximum uncertainty of the results. It is a case when all values of variable Ξ in the range $a\ldots b$ have the same probability. The corresponding probability density (called uniform) of such a random variable Ξ is:

$$f(\xi) = \begin{cases} \dfrac{1}{b-a}, & a \le \xi \le b, \\ 0, & \xi < 0, \xi > b. \end{cases}$$

A view of the uniform density probability is represented in Fig. 1.2.

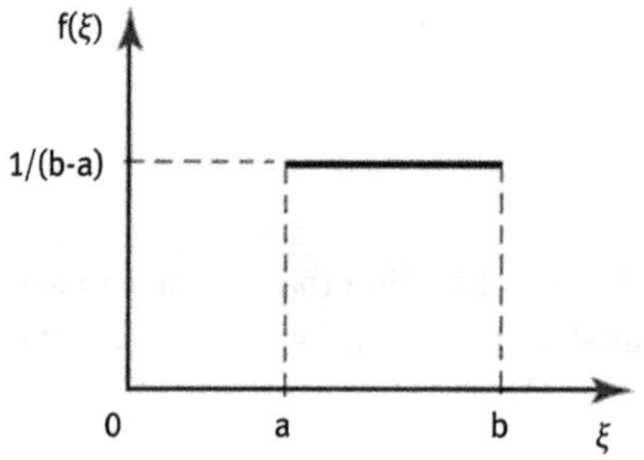

Fig. 1.2: Uniform *PDF.*

The uniformly distributed variable Ξ has the expectation

$$\mu_\xi = \int_a^b \xi \cdot f(\xi) = \frac{a+b}{2},$$

and variance

$$D_\xi = \int_a^b \left(\xi - \mu_\xi\right)^2 \cdot f(\xi) d\xi = (b-a)^2/12.$$

The uniform distribution has its merit when one is looking for maximum entropy of experimental results.

Another type of probability density under examination is called the normal (Gaussian) distribution. The distribution of the normal value is described by the Gauss law:

$$f(\xi) = \frac{1}{\sqrt{2\pi} \cdot \sigma_\xi} \exp\left[- \frac{(\xi - \mu_\xi)^2}{2D_\xi} \right].$$

A view of the *PDF* of the normal random value is represented in Fig. 1.3.

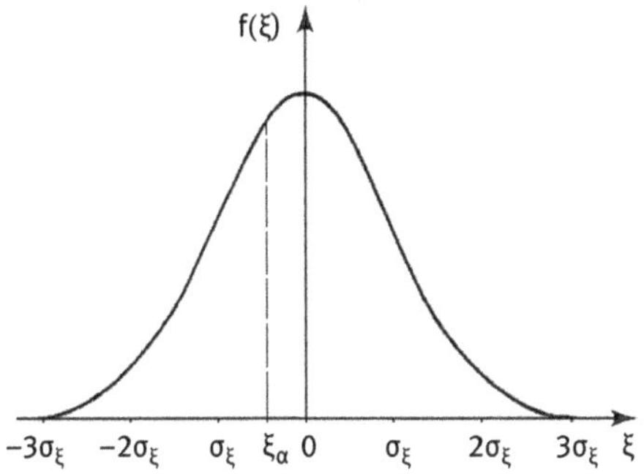

Fig. 1.3: The Gauss' *PDF*.

Here, ξ_α is the α-probability value of the random variable that meets the following condition:

$$P(\xi \le \xi_\alpha) = \int_{-\infty}^{\xi_\alpha} f(\xi) d\xi = \alpha.$$

In other words, this value determines the *PDF*'s segment located to the left of ξ_α whose area equals α.

The Gauss distribution function is completely defined by two moments: μ_ξ and D_ξ. In this case the expectation is a center of grouping of random variable values, with the variance being a measurement of their scattering around the expectation. When the variance is small, random variable values are grouped in the neighborhood of the expectation μ_ξ, and if σ_ξ is large, generally speaking, the values will be more spread around the mathematical expectation. One importance of Gaussian distribution in probability theory is based on the central limit theorem. Its engineering interpretation states that summation (action) of a large number of independent random values (factors) with similar distributions produces a random value (result) with a distribution tending to the normal distribution.

In technical applications, there is often a need to simultaneously consider a set of random variables characterized the operational status of the object. In this case it is more practical to use a random vector $\boldsymbol{\Xi}^{\mathrm{T}} = (\Xi_1, \ldots, \Xi_n)$ instead of several random variables $\Xi_1, \ldots, \Xi_n$. Symbol "T" denotes the transposition of the vector $\boldsymbol{\Xi}$:

$$\boldsymbol{\Xi} = \begin{pmatrix} \Xi_1 \\ \cdot \\ \cdot \\ \Xi_n \end{pmatrix}.$$

When a vector variable is used, one has to deal with a multivariate distribution function of the kind:

$$F\left(\boldsymbol{\xi}^{\mathrm{T}}\right) = F(\xi_1, \ldots, \xi_n) = F(\Xi_1 \le \xi_1, \ldots, \Xi_n \le \xi_n).$$

If function $F(\boldsymbol{\xi}^{\mathrm{T}})$ has partial derivatives with respect to ξ_i, the joint probability density of variables $\Xi_1, \ldots, \Xi_n$ has the form:

$$f\left(\boldsymbol{\xi}^{\mathrm{T}}\right) = f(\xi_1, \ldots, \xi_n) = \frac{\partial^n F(\xi_1, \ldots, \xi_n)}{\partial \xi_1 \ldots \partial \xi_n}.$$

Probability densities of the type:

$$f(\xi_i) = \int_{-\infty}^{\infty} \cdots \int_{-\infty}^{\infty} f(\xi_1, \ldots, \xi_n) d\xi_1 \ldots d\xi_{i-1} d\xi_{i+1} \ldots d\xi_n$$

are referred to as marginal.

Random variables $\Xi_1, \ldots, \Xi_n$ are called independent, if

$$f(\xi_1, \ldots, \xi_n) = f(\xi_1) \cdot \ldots \cdot f(\xi_n).$$

When variables are dependent, that is, the probability of Ξ_i depends on the remaining variables' magnitude, then:

$$f(\xi_1, \ldots, \xi_n) = f(\xi_i / \xi_1, \ldots, \xi_{i-1}, \xi_{i+1}, \ldots, \xi_n) \times f(\xi_1, \ldots, \xi_{i-1}, \xi_{i+1}, \ldots, \xi_n).$$

Here $f(\xi_i / \xi_1, \ldots, \xi_{i-1}, \xi_{i+1}, \ldots, \xi_n)$ is a conditional probability density determining the probability of an event $\xi_i < \Xi_i \le \xi_i + d\xi_i$ when values of remaining $(n - 1)$ variables are known.

A statistical relationship between variables Ξ_i and Ξ_j is characterized by the second-order moment called the cross-covariance (covariance):

$$\gamma_{ij} = M[(\Xi_i - \mu_i)(\Xi_j - \mu_j)] = M[(\Xi_j - \mu_j)(\Xi_i - \mu_i)].$$

As it follows from its definition, the covariance is positive if values $\Xi_i > \mu_i (\Xi_i < \mu_i)$ appear most often along with values $\Xi_j > \mu_j (\Xi_j < \mu_j)$, otherwise the covariance is negative. It is more convenient to quantify the relationships between variables through the use of the normalized version of the cross-covariance called the cross-correlation coefficient:

$$\rho_{ij} = \frac{\gamma_{ij}}{\sigma_i \sigma_j},$$

whose values are set to within $-1\ldots+1$ range. The range limits (values ±1) correspond to a linear dependence of the two variables; the correlation coefficient and covariance are null when the variables are independent.

Statistical relationships between n random variables are described by a covariance matrix of the form:

$$\mathbf{\Gamma}_\Xi = \begin{pmatrix} \gamma_{11} & \cdot & \gamma_{1n} \\ \cdot & \cdot & \cdot \\ \gamma_{n1} & \cdot & \gamma_{nn} \end{pmatrix}.$$

The normalized version of the matrix $\mathbf{\Gamma}$ is a correlation matrix:

$$\mathbf{P}_\Xi = \begin{pmatrix} 1 & \cdot & \rho_{1n} \\ \cdot & \cdot & \cdot \\ \rho_{n1} & \cdot & 1 \end{pmatrix},$$

where $\rho_{ii} = \frac{\gamma_{ii}}{\sigma_i \sigma_i} = 1$. From their definition, $\gamma_{ij} = \gamma_{ji}$ and $\rho_{ij} = \rho_{ji}$, so the covariance and correlation matrices are symmetric.

For example, consider n random Gauss-distributed variables whose probability density is given as follows:

$$f\left(\boldsymbol{\xi}^{\mathrm{T}}\right) = (2\pi)^{-n/2} |\mathrm{H}|^{1/2} \exp\left[-\frac{1}{2}(\boldsymbol{\xi}-\boldsymbol{\mu})^{\mathrm{T}} \mathbf{H} (\boldsymbol{\xi}-\boldsymbol{\mu})\right],$$

where $(\boldsymbol{\xi}-\boldsymbol{\mu})^{\mathrm{T}} = (\xi_1 - \mu_1, \ldots, \xi_n - \mu_n)$; $\mathbf{H}$ is a matrix inverse to the covariance matrix and is written as follows:

$$\mathbf{H} = \mathbf{\Gamma}^{-1} = \{\eta_{ij}\}, \quad i, j = 1, \ldots, n; \quad |\mathrm{H}| = |\Gamma|^{-1}.$$

Therefore, the exponent index in the expression of the probability density $f(\xi^{\mathrm{T}})$ may be represented in terms of matrix $\mathbf{H}$ elements:

$$(\boldsymbol{\xi}-\boldsymbol{\mu})^{\mathrm{T}} \mathbf{H} (\boldsymbol{\xi}-\boldsymbol{\mu}) = \sum_{i=1}^{n} \cdot \sum_{j=1}^{n} \eta_{ij} (\xi_i - \mu_i)(\xi_j - \mu_j).$$

In particular, for two random variables Ξ_1 and Ξ_2 the covariance matrix $\mathbf{\Gamma}$ and inverse matrix $\mathbf{H}$ are:

$$\mathbf{\Gamma} = \begin{pmatrix} \sigma_1^2 & \gamma_{12} \\ \gamma_{21} & \sigma_2^2 \end{pmatrix}; \qquad \mathbf{H} = \frac{1}{\sigma_1^2\sigma_2^2 - \gamma^2_{12}} \begin{pmatrix} \sigma_2^2 & -\gamma_{12} \\ -\gamma_{21} & \sigma_1^2 \end{pmatrix}.$$

The joint probability density of two variables is written as follows:

$$f(\xi_1, \xi_2) = \frac{1}{2\pi\sqrt{\sigma_1^2\sigma_2^2 - \gamma_{12}^2}} \times \exp\left[- \frac{\sigma_2^2(\xi_1 - \mu_1)^2 + \sigma_1^2(\xi_2 - \mu_2)^2 - 2\gamma_{12}(\xi_1 - \mu_1)(\xi_2 - \mu_2)}{2(\sigma_1^2\sigma_2^2 - \gamma_{12}^2)} \right].$$

The joint probability density of the bivariate Gaussian distribution is conditionally represented in Fig. 1.4.

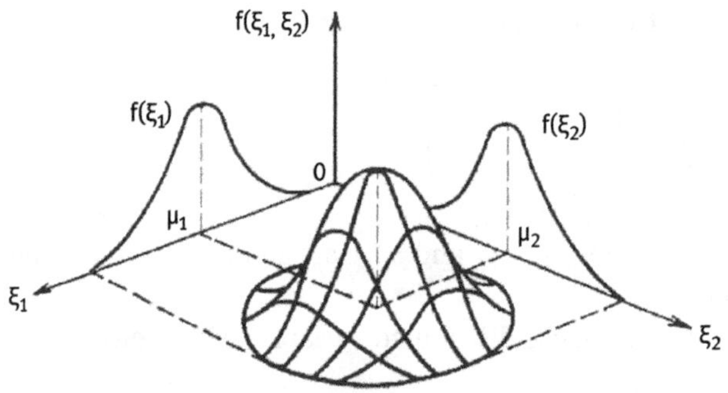

Fig. 1.4: The *PDF* of the bivariate Gauss distribution.

1.1.2 Function of random variables

As rule, test results are a product of processing measurement results. Because measurement results are random values, then it is logical to consider a function Ψ of random variables Ξ_i with μ_i and σ_i such that:

$$\Psi = \varphi(\Xi_1, \ldots, \Xi_i, \ldots, \Xi_n).$$

Clearly, values of Ψ will be random as well. First of all, one can examine the function Ψ in the form of a linear transformation:

$$\Psi = a \cdot \Xi + b,$$

where a and b are constant factors. In the sample space **A**, such a transformation is reduced only to the movement of subspace A_i (event) and its scaling. However, the placement (nature) of elements (points) of the space **A** does not change. In other words, the probabilistic scheme for constructing sample space **A** does not change. Therefore, the type of the distribution law of Ψ will corresponds to one that has a variable Ξ. However, the parameters of the *PDF* will be different:

$$\mu_\psi = \mu_\xi + b; \;\; \mathbf{D}_\psi = a^2 \mathbf{D}_\xi.$$

Consider the case when the function Ψ is non-linear. The probability of occurrence of a value ξ_i in a small interval $d\xi$ can be determined as $P_{d\xi} = f(\xi_i)d\xi$. The corresponding interval of the function Ψ will be equal to:

$$d\psi = \left|\frac{d\psi}{d\xi}\right|_{\xi_i} d\xi.$$

The derivative module is used because only sizes of intervals are considered.

Since the probabilities $P_{d\xi}$ and $P_{d\psi}$ must be the same, it follows from this that:

$$f(\psi_i)d\psi = f(\xi_i)d\xi,$$

where $\psi_i = \varphi(\xi_i)$. From the above equality it follows that the *PDF* of the *k*-valued function Ψ is defined by the following expression:

$$f(\psi) = kf(\xi)\left|\frac{d\xi}{d\psi}\right|.$$

Now, we turn to the case when Ψ is a sum of two independent variables Ξ_1 and Ξ_2:

$$\Psi = \Xi_1 + \Xi_2.$$

The distribution function of Ψ may be found in the following way:

$$F(\psi) = F(\Psi \le \psi) = P(\Xi_1 + \Xi_2 < \psi) = \iint\limits_{A} f(\xi_1, \xi_2) d\xi_1 d\xi_2,$$

with the integration region A represented in Fig. 1.5.

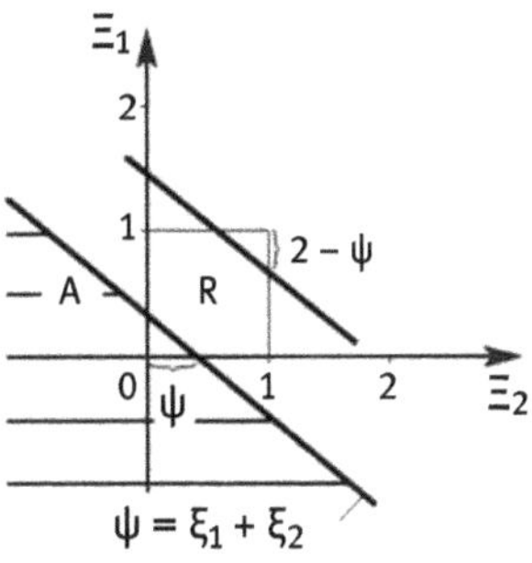

Fig. 1.5: The integration region A.

Due to the independence of variables Ξ_1 and Ξ_2, the joint distribution density $f(\xi_1, \xi_2)$ will be simply a product of the *PDFs* of these two variables:

$$f(\xi_1, \xi_2) = f(\xi_1) \cdot f(\xi_2).$$

Hence the expression of $F(\psi)$ may be transformed into the following expression:

$$F(\psi) = \iint_A f(\xi_1) \cdot f(\xi_2) d\xi_1 d\xi_2.$$

In the case when Ξ_1 and Ξ_2 have uniform distribution on interval of their values $0\ldots1$, the distribution function of Ψ will be equal to:

$$F(\psi) = \iint_A f(\xi_1) f(\xi_2) d\xi_1 d\xi_2 = \iint_{A \cap R} d\xi_1 d\xi_2.$$

That is, it equals to the area of the intersection of the region A and square R (Fig. 1.5). In other words, the value of $F(\psi)$ is an area of the zone of R located under the line $\psi = \xi_1 + \xi_2$. This area can be defined as follows:

a) $F(\psi) = 0, \ \psi \le 0;$

b) $F(\psi) = \psi^2/2, \ 0 < \psi \le 1;$

c) $F(\psi) = 1 - (2 - \psi)^2/2, \ 1 < \psi \le 2;$

d) $F(\psi) = 0, \ \psi > 0.$

Differentiation of $F(\psi)$ allows to determine the *PDF* in the following form:

a) $f(\psi) = 0, \ \psi \le 0;$

b) $f(\psi) = \psi, \ 0 < \psi \le 1;$

c) $f(\psi) = 2 - \psi, \ 1 < \psi \le 2;$

d) $f(\psi) = 0, \ \psi > 0.$

Therefore, $f(\psi)$ has a shape of an isosceles triangle with a base and height of 2 and 1, respectively. Such type of a distribution law is called symmetricly triangular and has parameters $\mu_\psi = 1$ and $D_\psi = 1/6$.

As it was mentioned earlier, the distribution function of a sum of two independent random variables may be transformed to the following form:

$$F(\psi) = \iint_A f(\xi_1) \cdot f(\xi_2) d\xi_1 \, d\xi_2$$

$$= \int_{-\infty}^{\infty} f(\xi_1) d\xi_1 \int_{-\infty}^{\psi - \xi_1} f(\xi_2) d\xi_2.$$

Corresponding probability density of the function Ψ will be equal to:

$$f(\psi) = \frac{dF(\psi)}{d\psi} = \int_{-\infty}^{\infty} f(\xi_1)\cdot f(\psi-\xi_1)d\xi_1.$$

Such a type of integral is called the convolution integral which can be noted as:

$$f(\psi) = f(\xi_1) * f(\xi_2),$$

and the corresponding mathematical operation is called a convolution. As an example, let us define the probability density of two independent variables with Gauss law using the convolution integral:

$$\begin{aligned} f(\psi) &= \int_{-\infty}^{\infty} f(\xi_1)\cdot f(\psi-\xi_1)d\xi_1 \\ &= \frac{1}{2\pi\sigma_1\sigma_2}\int_{-\infty}^{\infty} \exp\left[-\frac{(\xi_1-\mu_1)^2}{2\sigma_1^2} - \frac{(\psi-\xi_1-\mu_2)^2}{2\sigma_2^2}\right]d\xi_1 \\ &= \frac{1}{2\pi\sigma_1\sigma_2}\int_{-\infty}^{\infty} \exp(-A\xi_1^2 + 2B\xi_1 - C)d\xi_1, \end{aligned}$$

where

$$A = \frac{(\sigma_1^2+\sigma_2^2)}{2\sigma_1\sigma_2},\ B = \frac{\mu_1}{2\sigma_1^2} + \frac{\psi-\mu_2}{2\sigma_2^2},\ C = \frac{\mu_1^2}{2\sigma_1^2} + \frac{(\psi-\mu_2)^2}{2\sigma_2^2}.$$

A solution to the integral of an exponential function is known to be as follows

$$\int_{-\infty}^{\infty} \exp(-Ax^2 \pm 2Bx - C)dx = \sqrt{\frac{\pi}{A}}\exp\left(-\frac{AC+B^2}{A}\right).$$

Taking this into account, the *PDF* may be written as:

$$f(\psi) = \frac{1}{\sqrt{2\pi(\sigma_1^2+\sigma_2^2)}}\cdot\exp\left\{-\frac{[\psi-(\mu_1+\mu_2)]^2}{2(\sigma_1^2+\sigma_2^2)}\right\}.$$

Therefore, the function $\Psi = \Xi_1 + \Xi_2$ has a Gaussian distribution with the expectation:

$$\mu_\psi = \mu_1 + \mu_2,$$

and the square root of the variance:

$$\sigma_\psi = \sqrt{D_\psi} = \sqrt{\sigma_1^2 + \sigma_2^2}.$$

If the function Ψ is equal to:

$$\Psi = a_1\Xi_1 + a_2\Xi_2,$$

its first and second moments are

$$\mu_\psi = a_1\mu_1 + a_2\mu_2;\ \sigma_\psi = \sqrt{a_1^2\sigma_1^2 + a_2^2\sigma_2^2}.$$

In case of dependent variables, their sum still follows the Gaussian law with the same expectation, but the variance will include the cross-correlation of variables Ξ_1 and Ξ_2:

$$\begin{aligned} D_\psi &= M[(\psi - \mu_\psi)^2] = M[(\xi_1 + \xi_2 - \mu_1 - \mu_2)^2] \\ &= M[(\xi_1 - \mu_1)^2 + (\xi_2 - \mu_2)^2 + 2(\xi_1 - \mu_1)(\xi_2 - \mu_2)] \\ &= D_1 + D_2 + \gamma_{12} = \sigma_1^2 + \sigma_2^2 + 2\rho_{12}\sigma_1\sigma_2. \end{aligned}$$

If Ψ has a number of arguments $n > 2$ and is not a linear function, it may be linearized at the point $\boldsymbol{\mu}^\mathrm{T} = (\mu_1, \ldots, \mu_i, \ldots, \mu_n)$ yielding the following expression:

$$\Psi \approx \varphi(\mu_1, \ldots, \mu_n) + \sum_{i=1}^{n} a_i(\Xi_i - \mu_i)$$

where $a_i = \frac{\partial\psi}{\partial\xi_i}|_\mu$. We proceed to consider the variable E with the probabilistic properties equal to Ψ:

$$\mathrm{E} = \Psi - \varphi(\mu_1, \ldots, \mu_i, \ldots, \mu_n) = \sum_{i=1}^{n} a_i(\Xi_i - \mu_i) = \sum_{i=1}^{n} a_i Z_i.$$

The variable Z_i has parameters $\mu_\zeta = 0$ and D_i.

Commutative and associative properties of the convolution integral in the following forms:

$$[f(\zeta_1) * f(\zeta_2)] * \ldots * f(\zeta_n) = f(\zeta_1) * \ldots * [f(\zeta_{n-1}) * f(\zeta_n)]$$

and

$$f(\zeta_1) * f(\zeta_2) * \ldots * f(\zeta_n) = f(\zeta_2) * f(\zeta_1) * \ldots * f(\zeta_n),$$

allow to determine $f(\psi)$ via the following step-by-step approach:

$$\begin{aligned} \mathrm{E}_1 &= a_1 Z_1 + a_2 Z_2, \\ \mathrm{E}_2 &= \mathrm{E}_1 + a_3 Z_3, \\ &\cdot \\ \mathrm{E} &= \mathrm{E}_{n-1} + a_n Z_n. \end{aligned}$$

In a case where Z_i is a normal variable, the function E will have the normal distribution with

$$\mu_\varepsilon = 0;$$

$$D_\varepsilon = \sum_{i=1}^{n} a_i^2 D_i + 2\sum_{i \neq j} a_i a_j \gamma_{ij}.$$

Accordingly, the function Ψ will have the normal distribution as well with parameters $\mu_\psi = \varphi(\mu_1, \ldots, \mu_i, \ldots, \mu_n)$ and $D_\psi = D_\varepsilon$.

Very often, data processing may be represented as a system of linear equations:

$$\Psi_1 = \sum_{i=1}^{n} a_{1i}\Xi_i + b_1,$$
$$\cdot$$
$$\Psi_k = \sum_{i=1}^{n} a_{ki}\Xi_i + b_k,$$

which may be written in the matrix form as:

$$\boldsymbol{\Psi} = \boldsymbol{A} \cdot \boldsymbol{\Xi} + \boldsymbol{B},$$

where

$$\boldsymbol{\Psi} = \begin{pmatrix} \Psi_1 \\ \cdot \\ \Psi_k \end{pmatrix}, \quad \boldsymbol{\Xi} = \begin{pmatrix} \Xi_1 \\ \cdot \\ \Xi_n \end{pmatrix},$$

$$\boldsymbol{A} = \begin{pmatrix} a_{11} & \cdot & a_{1n} \\ \cdot & \cdot & \cdot \\ a_{k1} & \cdot & a_{kn} \end{pmatrix}, \quad \boldsymbol{B} = \begin{pmatrix} b_1 \\ \cdot \\ b_k \end{pmatrix}.$$

The mathematical expectation of vector $\boldsymbol{\Psi}$ is:

$$M[\boldsymbol{\Psi}] = \boldsymbol{\mu}_{\boldsymbol{\Psi}} = \boldsymbol{A}\boldsymbol{\mu}_{\boldsymbol{\Xi}} + \boldsymbol{B},$$

and its covariance matrix has the following form:

$$\boldsymbol{\Gamma}_{\boldsymbol{\Psi}} = \begin{pmatrix} D_1 & \cdot & \gamma_{1k} \\ \cdot & \cdot & \cdot \\ \gamma_{k1} & \cdot & D_k \end{pmatrix}.$$

The elements of the $\boldsymbol{\Gamma}_{\boldsymbol{\Psi}}$ may be interpreted as the mathematical expectation of ij^{th} element being a product of vector $(\boldsymbol{\Psi} - \boldsymbol{\mu}_{\boldsymbol{\Psi}})$ and $(\boldsymbol{\Psi} - \boldsymbol{\mu}_{\boldsymbol{\Psi}})^{\mathrm{T}}$:

$$(\boldsymbol{\Psi} - \boldsymbol{\mu}_{\boldsymbol{\Psi}}) = \begin{pmatrix} \Psi_1 - \mu_1 \\ \Psi_2 - \mu_2 \\ \cdot \\ \Psi_k - \mu_k \end{pmatrix}.$$

Such an approach allows to represent the matrix $\mathbf{\Gamma}_\psi$ as follows:

$$\begin{aligned}
\mathbf{\Gamma}_\psi &= M\left[(\boldsymbol{\Psi}-\boldsymbol{\mu}_\Psi)(\boldsymbol{\Psi}-\boldsymbol{\mu}_\Psi)^{\mathrm{T}}\right] \\
&= M\left[(\boldsymbol{A}\Xi+\boldsymbol{B}-\boldsymbol{A}\boldsymbol{\mu}_\Xi-\boldsymbol{B})(\boldsymbol{A}\Xi+\boldsymbol{B}-\boldsymbol{A}\boldsymbol{\mu}_\Xi-\boldsymbol{B})^{\mathrm{T}}\right] \\
&= M\left[\boldsymbol{A}\cdot(\Xi-\boldsymbol{\mu}_\Xi)\cdot(\Xi-\boldsymbol{\mu}_\Xi)^{\mathrm{T}}\cdot\boldsymbol{A}^{\mathrm{T}}\right] \\
&= \boldsymbol{A}\cdot M\left[(\Xi-\boldsymbol{\mu}_\Xi)(\Xi-\boldsymbol{\mu}_\Xi)^{\mathrm{T}}\right]\cdot\boldsymbol{A}^{\mathrm{T}} = \boldsymbol{A}\mathbf{\Gamma}_\Xi\boldsymbol{A}^{\mathrm{T}}.
\end{aligned}$$

Therefore, the knowledge of the covariance matrix of variables Ξ allows to calculate the covariance matrix of the results of processing.

1.2 Adaptation of probability models

This section deals with the elements of mathematical statistics which comprise problems related to adaptation of probability models to experimental data. Such data is represented by a sequence of real numbers (series) $x_1,\ldots,x_n$ of a variable X. All x_i are assume to be independent random values with an identical $f(x)$, so the series is characterized by the following probability density:

$$f(x_1,\ldots,x_n) = \prod_{i=1}^{n} f(x_i).$$

An adaptation procedure is realized by utilizing a probability model as a working hypothesis and its subsequent matching with experimental data. This procedure has a random character due to utilizing a limited number of x_i from an infinite population of random value X. Methods of mathematical statistics give instructions regarding better approaches to utilize experimental data as well as to evaluate reliability of inferences pertaining to the adapted models.

1.2.1 Processing of experimental data

Processing series $x_1,\ldots,x_n$ and presentation of its results in a view suitable for inference making are critical issues of statistical analysis. Usually, it starts with a preliminary processing of experimental data which reduces the latter to a few statistics. Under the name of statistic T_n is understood to be a result of processing the series $x_1,\ldots,x_n$:

$$T_n = T(x_1,\ldots,x_n).$$

There are a few widely used statistics in engineering practice. Random variable Z, called the standard normal variable, plays a basic role:

$$Z = (X - \mu_x)/\sigma_x,$$

where X is a random variable with the standard normal distribution. It is an obvious fact that Z has the mathematical expectation $\mu_z = 0$ and $\sigma_z = 1$.

The statistic T_n in the form:

$$T_n = \sum_{i=1}^{n} z_i^2 = \chi_n^2$$

is called as chi-squared variable with n degrees of freedom. In general, the degrees of freedom of a statistic equal to the number of independent components used in its calculation. A mathematical description of the χ_n^2 *PDF* is:

$$f(\chi_n^2) = \begin{cases} \frac{(\chi_n^2)^{\frac{n}{2}-1} \cdot \exp(-\chi_n^2)}{2^{n/2} \cdot \Gamma(n/2)}, & \chi_n^2 > 0, \\ 0, & \text{otherwise} \end{cases}$$

where $\Gamma(n/2)$ is the gamma function whose value (for integer $n \geq 2$) coincides with the factorial $(n-1)!$; n – the number of degrees of freedom. Plots of $f(\chi_n^2)$ are represented in Fig. 1.6.

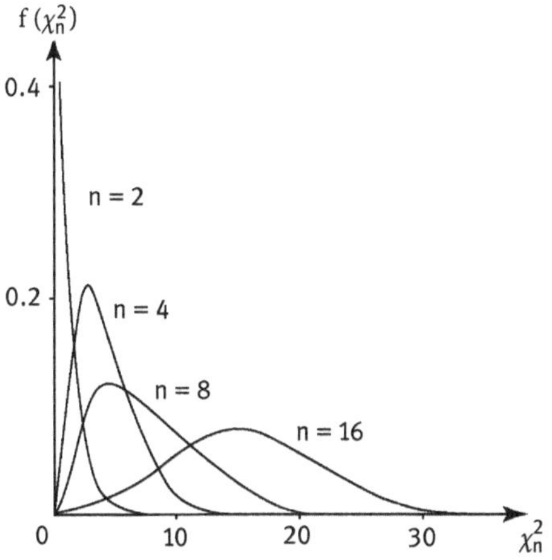

Fig. 1.6: The *PDF* of the chi-squared variable.

The first two moments of the χ_n^2 distribution are $\mu_{\chi_n^2} = n$, and $D_{\chi_n^2} = 2n$.

The χ_n^2 distribution relates to a sum of independent variables z_i^2 with the same variance. This distribution approaches the Gauss law approximation when n increases due to the central limit theorem. In particular, if $n > 30$, a variable $\sqrt{2\chi_n^2}$ has a distribution similar to normal with $\mu_{\chi_n^2} = \sqrt{2n-1}$, and $D_{\chi_n^2} = 1$.

Statistic

$$T_n = \frac{Z}{\sqrt{\chi_n^2/n}} = t_n$$

is a variable with Student's *t*-distribution. Its *PDF* is given by the following expression:

$$f(t_n) = \frac{\Gamma\left(\frac{n+1}{2}\right)}{\sqrt{\pi n}\cdot\Gamma(n/2)}\cdot\left(1+\frac{t_n^2}{n}\right)^{-(n+1)/2},$$

where n is the number of degrees of freedom (Fig. 1.7).

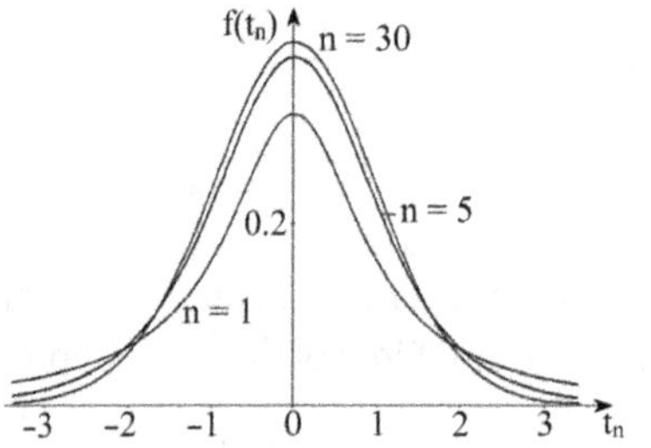

Fig. 1.7: The Student's *PDF*.

The characteristics of this distribution are $\mu_{t_n}=0$, and $D_{t_n}=n/(n-2)$ for $n>2$. Asymptotically ($n\to\infty$), the *t*-distributed variable approaches the standard normal variable *Z*. Actually, a good approximation can be obtained with $n>30$.

The final statistic under examination is

$$T_n = \frac{\chi_{n_1}^2\cdot n_2}{\chi_{n_2}^2\cdot n_1} = F_{n_1,n_2}.$$

This statistic is called the *F*(Fisher-Snedecor)-statistic with n_1 and n_2 degrees of freedom. A view of its *DPF* described by the expression:

$$f(F_{n_1,n_2}) = \frac{\Gamma\left(\frac{n_1+n_2}{2}\right)\left(\frac{n_1}{n_2}\right)^{n_1/2}\cdot F_{n_1,n_2}{}^{(n_1/2)-1}}{\Gamma\left(\frac{n_1}{2}\right)\Gamma\left(\frac{n_2}{2}\right)\left(1+\frac{n_1}{n_2}F_{n_1,n_2}\right)^{(n_1+n_2)/2}},\quad F_{n_1,n_2}\geq 0,$$

is represented in Fig. 1.8.

Parameters of the *F*-statistic are:

$$\mu_F = \frac{n_2}{n_2-2},\ \sigma_F^2 = \frac{2n_2^2(n_1+n_2-2)}{n_1(n_2-2)^2(n_2-4)};\ n_2>4.$$

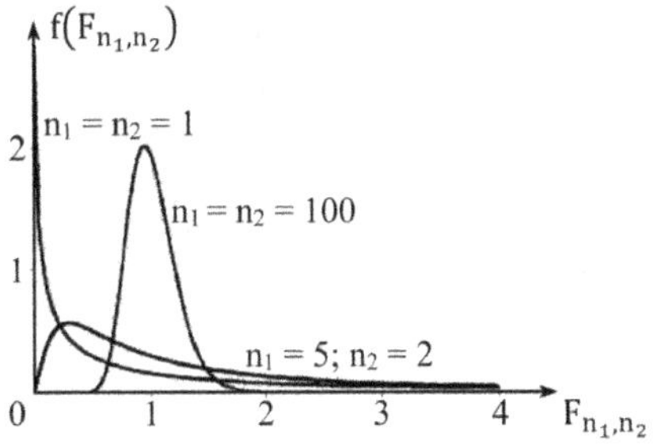

Fig. 1.8: The *PDF* of the *F*-statistic.

Asymptotically $(n_1, n_2 \to \infty)$, the variable F approaches the normal distribution with $\mu_F = 1$, and $\sigma_F^2 = \frac{2(n_1+n_2)}{n_1 n_2}$. An acceptable ($\approx 10\%$) approximation is achieved with $n_1, n_2 > 60$.

1.2.2 Criterion of maximum likelihood

The task of model factor estimation can be stated as follows: there is a random variable X with a known probability density $f(x/\boldsymbol{a})$ featuring parameters $\boldsymbol{a}^{\mathrm{T}} = (a_1, \ldots, a_k)$. Existing values of these parameters provided observations $\boldsymbol{x}^{\mathrm{T}} = (x_1, \ldots, x_n)$ are unknown although their values are fixed. It is required to estimate parameters $\boldsymbol{a}$ to consider that some of their values "most likely" provide observations $\boldsymbol{x}$.

In 1925, R.A. Fisher, a British statistician and geneticist, has formulated the maximum likelihood criterion which holds a key position in the statistic estimation theory. Its essence may be described in the following manner.

Before being observed, the possible values of random variable X are described by the probability density $f(\boldsymbol{x}/\boldsymbol{a})$. Once observations $x_1, \ldots, x_n$ are obtained, it is appropriate to proceed with considering the possible values of parameters $\boldsymbol{a}$ that provided those observations. To do this one constructs a likelihood function $L(\boldsymbol{a}/\boldsymbol{x})$ which is proportional to $f(\boldsymbol{x}/\boldsymbol{a})$ with observations $\boldsymbol{x}$ and unknown values of $\boldsymbol{a}$. The maximum likelihood estimates (*MLE*s) of parameters $\boldsymbol{a}$ designated as $\tilde{\boldsymbol{a}}$ provide the maximum value of the function $L(\boldsymbol{a}/\boldsymbol{x})$:

$$L(\tilde{\boldsymbol{a}}/\boldsymbol{x}) = \max_{\tilde{\boldsymbol{a}}} L(\boldsymbol{a}/\boldsymbol{x})$$

In other words, the *MLE*s correspond to the highest probability of appearance of observations $x_1, \ldots, x_n$. The likelihood function plays a fundamental role in the estimation theory since it carries all information about model parameters obtainable from experimental data [1].

Often it is more convenient to use the logarithmic likelihood function

$$l(\boldsymbol{a}/\boldsymbol{x}) = \ln L(\boldsymbol{a}/\boldsymbol{x}),$$

which comprises additive constants instead of proportionality coefficients. Furthermore, $l(\boldsymbol{a}/\boldsymbol{x})$ takes on a simple view for exponential forms of distribution laws which are the most often case in technical applications.

To illustrate the principle of maximum likelihood let us take an example of estimating parameters for the *PDF* of a normal variable X with μ_x and D_x. The results of this variable measurement are considered as a vector $\boldsymbol{x}$ of independent readouts $x_1, \ldots, x_n$ with the joint probability density:

$$f(x_1, \ldots, x_n) = (2\pi D_x)^{-n/2} \cdot \exp\left\{ - \sum_{i=1}^{n} \frac{(x_i - \mu_x)^2}{2D_x} \right\}.$$

As it was mentioned previously, the likelihood function is derived from the probability density providing fixed x_i and variable parameters μ_x and D_x. In particular, the logarithmic likelihood function will have the form:

$$l(\mu_x, D_x/\boldsymbol{x}) = -\frac{n}{2} ln(2\pi D_x) - \frac{1}{2D_x} \sum_{i=1}^{n} (x_i - \mu_x)^2.$$

The conditions of a maximum of the logarithmic likelihood function with respect to the parameter μ_x will be:

$$\frac{\partial l}{\partial \mu_x} = \frac{1}{2D_x} \sum_{i=1}^{n} 2(x_i - \mu_x) = 0,$$

which will result in the following expression:

$$\tilde{\mu}_x = \frac{\sum_{i=1}^{n} x_i}{n}. \tag{1}$$

Now, let's represent the condition of a maximum of the logarithmic likelihood function with respect to the parameter D_x:

$$\frac{\partial l}{\partial D_x} = -\frac{n}{2D_x} + \frac{1}{2D_x^2} \sum_{i=1}^{n} (x_i - \mu_x)^2 = 0.$$

This equation leads to the following expression for caclculating the *MLE* of a variance:

$$\tilde{D}_x = \tilde{\sigma}_x^2 = \frac{\sum_{i=1}^{n} (x_i - \mu_x)^2}{n}.$$

In addition to the point estimates of the parameters of general populations, interval estimates are also used in statistics. One of these interval estimates is the confidence interval, which is obtained on a basis of statistics computed from observed data. The confidence interval will contain the true value of the unknown estimated parameter of the general population with a probability that is specified by a given confidence level. Usually, a confidence level $P = 0.95$ is utilized. The practical interpretation of the confidence interval with a confidence level, say $P = 0.95$, is as follows. Let us assume a very large number of independent experiments with a similar construction of a confidence interval. Then in 95% of experiments the confidence interval will contain the true value of the estimated parameter. In the remaining 5% of experiments the confidence interval may not contain that value.

1.2.3 Properties of maximum likelihood estimates

A wide applications of the maximum likelihood criterion is based on a property of *MLE* invariance: if $\tilde{a}$ is the *MLE* of parameter a, then $g(\tilde{a})$ will be the *MLE* of any function of parameter a (not necessarily a one-to-one function) [1].

To demonstrate that property, let us examine an injective (one-to-one) function $g(a)$ instead of parameter a. A derivative of this function may be presented in the following form:

$$\frac{\partial l}{\partial a} = \frac{\partial l}{\partial g(a)} \cdot \frac{\partial g(a)}{\partial a}.$$

Respectively, the condition of a stationary point (extremum) of the logarithmic likelihood function

$$\frac{\partial l}{\partial g(a)} = 0$$

corresponds to the condition $\frac{\partial l}{\partial a} = 0$, if $\frac{\partial g(a)}{\partial a} \neq 0$.

The invariance property of the *MLE* is particularly important for estimation of model factors. This property substantially simplifies the calculation of *MLEs* of these factors when the relationship between the model factors and the statistical characteristics of the series are known. In this case, it suffices to obtain only the *MLEs* of these characteristics. Subsequently, the knowledge of a functional link between the statistical characteristics and model factors allows to calculate the *MLEs* of the latter.

In themselves, the parameter estimates cannot be correct or incorrect because they are somewhat arbitrary. Nevertheless, some estimates could be considered as "better" than others. To compare them, one can make use of the mean square error of $\tilde{a}$:

$$M[(\tilde{a} - a)^2] = M\{[\tilde{a} - M(\tilde{a})]^2\} + M\{[M(\tilde{a}) - a]^2\}.$$

The first term on the right-hand side of this expression is the estimate variance which is a measure of a "random" fraction of the error:

$$D_{\tilde{a}} = M\{[\tilde{a} - M(\tilde{a})]^2\}.$$

The second term – square of the estimate bias – gives a systematic deviation:

$$b_{\tilde{a}}^2 = M\{[M(\tilde{a}) - a]^2\}.$$

Depending on properties of components of the error, estimates come subdivided into several categories.

First, if the estimate expectation equals the parameter which is to be estimated, such as:

$$M(\tilde{a}) = a,$$

that is, $b_{\tilde{a}} = 0$, then the estimate is called unbiased.

Second, if the variance of the estimate $\hat{a}$ is less than that of any other estimate $\tilde{a}$, that is:

$$D_{\hat{a}} < D_{\tilde{a}},$$

then the estimate $\hat{a}$ is referred to as an efficient estimate.

And finally, if with the increase of the series size n the estimate draws near the parameter a with a probability tending to 1, in other words, at any small $c > 0$

$$\lim_{n\to\infty} P(|\tilde{a} - a| \geq c) = 0,$$

then the estimate is called consistent. From the Chebyshev inequality of the form:

$$P(|\tilde{a} - a| \geq c) = \frac{D_{\tilde{a}}}{c^2},$$

it follows that a sufficient (but not required) condition of consistency is:

$$\lim_{n\to\infty} D_{\tilde{a}} = 0.$$

In other words, the accuracy of the estimate must increase with an corresponding increase of n. Both conditions of the estimate's consistency are, in fact, requirements for the convergence in probability and the mean square.

Let us examine the properties of estimates of $\tilde{\mu}_x$ and $\tilde{D}_x$ deduced in the previous section. First of all, we'll evaluate an estimate of a mathematical expectation of $\tilde{\mu}_x$:

$$M[\tilde{\mu}_x] = \frac{1}{n}\sum_{i=1}^{n} M[x_i] = \frac{n}{n} \cdot \mu_x = \mu_x,$$

that is, this estimate is unbiased. An inherence about a consistency of estimates requires evaluating their variances. A variance of the estimate $\tilde{\mu}_x$ may be determined as:

$$D_{\tilde{\mu}} = Var\left[\frac{\sum_{i=1}^{n} x_i}{n}\right] = \frac{\sum_{i=1}^{n} Var[x_i]}{n^2} = \frac{nD_x}{n^2} = \frac{D_x}{n}.$$

A presence of the denominator n in this expression will decrease the variance $D_{\tilde{\mu}}$ with increasing n that is, the estimate of the μ_x is consistent.

In the case of an unknown expectation μ_x, its estimates $\tilde{\mu}_x$ may be used for calculation of the $\tilde{D}_x$:

$$\tilde{D}_x = \frac{\sum_{i=1}^{n} (x_i - \tilde{\mu}_x)^2}{n}.$$

First of all, we'll modify the numerator of the fraction:

$$\begin{aligned}
\sum_{i=1}^{n} (x_i - \tilde{\mu}_x)^2 &= (x_i - \mu_x + \mu_x - \tilde{\mu}_x)^2 \\
&= \sum_{i=1}^{n} (x_i - \mu_x)^2 - 2(\tilde{\mu}_x - \mu_x) \sum_{i=1}^{n} (x_i - \mu_x) + \sum_{i=1}^{n} (\tilde{\mu}_x - \mu_x)^2 \\
&= \sum_{i=1}^{n} (x_i - \mu_x)^2 + 2(\tilde{\mu}_x - \mu_x) \left(\sum_{i=1}^{n} x_i - n\mu_x \right) + n(\tilde{\mu}_x - \mu_x)^2 \\
&= \sum_{i=1}^{n} (x_i - \mu_x)^2 - 2(\tilde{\mu}_x - \mu_x) n(\tilde{\mu}_x - \mu_x) + n(\tilde{\mu}_x - \mu_x)^2 \\
&= \sum_{i=1}^{n} (x_i - \mu_x)^2 - n(\tilde{\mu}_x - \mu_x)^2.
\end{aligned}$$

The use of this expression allows to determe the mathematical expectation of the $\tilde{D}_x$ as follows:

$$\begin{aligned}
M[\tilde{D}_x] &= \frac{1}{n} M\left[\sum_{i=1}^{n} (x_i - \mu_x)^2\right] = \frac{1}{n}\left\{ M\left[\sum_{i=1}^{n} (x_i - \mu_x)^2\right] - M[n(\tilde{\mu}_x - \mu_x)^2] \right\} \\
&= \frac{1}{n}(nD_x - nD_{\tilde{\mu}}) = \frac{1}{n}(nD_x - D_x) = \frac{n-1}{n} D_x.
\end{aligned}$$

Therefore, the unbiased estimate $\tilde{D}_x$ has to be calculated in accordance with the following expression:

$$\tilde{D}_x = \frac{\sum_{i=1}^{n} (x_i - \tilde{\mu}_x)^2}{n-1}. \tag{2}$$

The variance of this estimate may be represented as:

$$D_{\tilde{D}_x} = Var\left[\frac{\sum_{i=1}^{n} (x_i - \tilde{\mu}_x)^2}{(n-1)}\right] = \frac{\sum_{i=1}^{n} Var\left[(x_i - \tilde{\mu}_x)^2\right]}{(n-1)^2}.$$

It is clear that the numerator of this fraction has an order of magnitude n, while the denominator – n^2. Therefore, the value of $D_{\tilde{D}_x}$ will decrease with increasing n, which indicates the consistency of $\tilde{D}_x$.

An inference about the efficiency of the *MLEs* can be obtained in the following fashion. The lower bound of the variance of an unbiased estimate of any parameter a is determined by the Cramer-Rao inequality:

$$D_{\hat{a}} \geq \frac{1}{I(a)}.$$

The $I(a)$ is called the Fisher information (or just information) contained in the series $x_1, \ldots, x_n$ about an unknown parameter a. In the case of *MLE*s, the Fisher information value is determined by the following expression:

$$I(a) = -M\left[\frac{\partial^2 l(a/(x_1, \ldots, x_n))}{\partial a^2}\right] = -M\left[l''\right]. \tag{3}$$

The efficient estimates have values corresponding to the lower bound of the Cramer-Rao inequality:

$$D_{\hat{a}} = \frac{1}{I(a)}.$$

Let us consider the value of l'. Due to existence of the second derivative of the likelihood function, l' may be linearized at a point $\hat{a}$:

$$l' = l'(\hat{a}) + (a - \hat{a}) \cdot l'' + \cdots.$$

Taking into account that $l'(\hat{a}) = 0$ for the *MLE* and ignoring the highest derivatives, the following expression can be written:

$$l' = (a - \hat{a}) \cdot l''.$$

Asymptotically, $l'' \underset{n\to\infty}{=} M\left[l''\right]$, so taking into account (3) the above written expression transforms to the following:

$$l' \underset{n\to\infty}{=} -(a - \hat{a}) \cdot I(\hat{a}) = -\frac{(a - \hat{a})}{D_{\hat{a}}}.$$

An integration of this expression yields the following form of the logarithmic likelihood function:

$$l(a) = -\frac{(a - \hat{a})^2}{2D_{\hat{a}}} + const.$$

Therefore, the likelihood function will correspond to the normal distribution

$$L(a) = C \cdot \exp\left[-\frac{(a-\hat{a})^2}{2D_{\hat{a}}}\right]$$

with the expectation $\hat{a}$ and variance $D_{\hat{a}} = 1/I(a)$. As this conclusion was achieved for $n \to \infty$, all *MLEs* are efficient asymptotically.

1.2.4 Least-squares method

The least-squares method is referred to the works of Adrien-Marie Legendre and Carl Gauss. Conceptually, it can be interpereted easily as follows:

"Results of n repeated and independent measurements $\tilde{x}_i$ may be represented as a sum of an unknown value x and a measurement error ξ_i, that is, $\tilde{x}_i = x + \xi_i$. Value x is estimated in such a way as to minimize the sum of squares of the error:

$$\sum_{i=1}^{n} \xi_i^2 = \sum_{i=1}^{n} (\tilde{x}_i - x)^2 \underset{x}{\Rightarrow} \min\text{''}.$$

The corresponding estimate of value x is called the least squares estimate (*LSE*), and it will be marked in the following text as $\hat{x}$.

Let us examine the results of measurements $\tilde{x}_i = x + \xi_i$ featuring $\mu_\xi = 0$ and D_ξ. As it was demonstrated in the previous section, asymptotically ($n \to \infty$), the likelihood function can be represented via fixed values of $\tilde{x}_i$ and an unknown variable x in the following form:

$$L(x) = C \cdot \exp\left[-\sum_{i=1}^{n} \frac{(\tilde{x}_i - x)^2}{2D_\xi}\right].$$

Therefore, the logarithmic likelihood function will be:

$$l(x) = const - \frac{1}{2D_x} \sum_{i=1}^{n} (\tilde{x}_i - x)^2.$$

If $n \to \infty$, the maximum of the likelihood function corresponds to the minimum of the sum $S(x) = \sum_{i=1}^{n} (\tilde{x}_i - x)^2$:

$$\max_x l(x) \Rightarrow \min_x S(x).$$

An extreme value of $S(x)$ is determined from a condition:

$$\frac{\partial S(x)}{\partial x} = 0,$$

which leads to the following equation:

$$2\sum_{i=1}^{n}(\tilde{x}_i - x) = 0.$$

Thus, the *MLE* of a value x is asymptotically equal to its *LSE*:

$$\hat{x} = \frac{\sum_{i=1}^{n}\tilde{x}_i}{n}.$$

The most popular applications of the least-squares method are related to approximating experimental data $\tilde{x}$ in a form of an analytical model

$$x = g(y, a_j).$$

Here a_j is a model factor $(j = 1,\ k)$, and y is an independent variable which values are known with zero or negligible errors.

The measurement result of x_i for a known value y_i may be presented as:

$$\tilde{x}_i = x_i + \xi_i = g(y_i, a_j) + \xi_i.$$

In accordance with the least-squares approach, the estimates of factors a_j can be found from the following condition:

$$S(a_j) = \sum_{i=1}^{n}\xi_i^2 = \sum_{i=1}^{n}[\tilde{x}_i - g(y_i, a_j)]^2 \underset{a_j}{\Rightarrow} \min.$$

Such a condition is determined as

$$\frac{\partial S(a_j)}{\partial a_j} = 0$$

which leads to a system of k equations:

$$\sum_{i=1}^{n}[\tilde{x}_i - g(y_i, a_j)] \cdot \frac{\partial g(y_i, a_j)}{\partial a_j} = 0,\ j = 1,\ k.$$

The solution to this system provides the *LSE*s of each factor a_j. As an example, approximating the measurements $\tilde{x}_i$ (signs "$*$" in Fig. 1.9) with a straight line may be examined.

In the linear case, the model $x = g\ (y, a_j)$ is represented by the following equation:

$$x = a_0 + a_1 y.$$

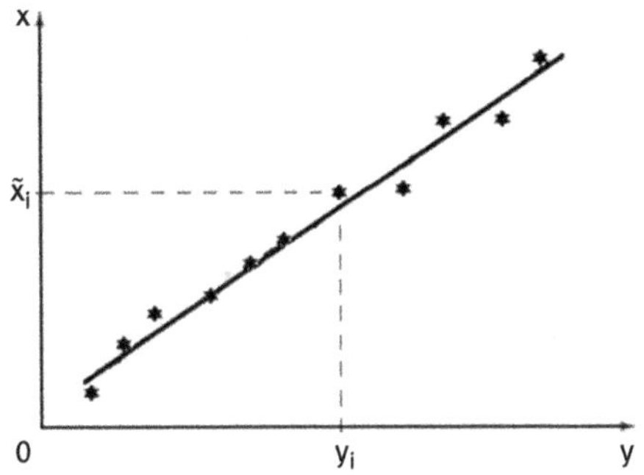

Fig. 1.9: Fitting measurement results.

Therefore, a system (3) providing the estimates $\hat{a}_j$is:

$$\sum_{i=1}^{n}(\tilde{x}_i - a_1 y_i - a_0) = 0;$$

$$\sum_{i=1}^{n}(\tilde{x}_i - a_1 y_i - a_0) y_i = 0;$$

These equations may be modified to the following form:

$$a_1 \sum_{i=1}^{n} y_i + n a_0 = \sum_{i=1}^{n} \tilde{x}_i;$$

$$a_1 \sum_{i=1}^{n} y_i^2 + a_0 \sum_{i=1}^{n} y_i = \sum_{i=1}^{n} (\tilde{x}_i \cdot y_i).$$

Using Cramer's rule provides the estimates of model factors:

$$\hat{a}_0 = \frac{\sum_{i=1}^{n} y_i^2 \cdot \sum_{i=1}^{n} \tilde{x}_i - \sum_{i=1}^{n} y_i \cdot \sum_{i=1}^{n} (\tilde{x}_i \cdot y_i)}{n \sum_{i=1}^{n} y_i^2 - \left(\sum_{i=1}^{n} y_i\right)^2};$$

$$\hat{a}_1 = \frac{n \sum_{i=1}^{n} (\tilde{x}_i \cdot y_i) - \sum_{i=1}^{n} y_i \cdot \sum_{i=1}^{n} \tilde{x}_i}{n \sum_{i=1}^{n} y_i^2 - \left(\sum_{i=1}^{n} y_i\right)^2}.$$

In fact, an estimate of the residual variance $D_{\tilde{x}}$

$$\hat{D}_{\tilde{x}} = \sum_{i-1}^{n} \frac{(\tilde{x}_i - \hat{a}_1 y_i - \hat{a}_0)^2}{n-2},$$

characterizes the scattering of the measurement resul. A number of degrees of freedom is $(n - 2)$ because two estimates $\hat{a}_0$ and $\hat{a}_1$ were already calculated from the realization $x_1, \ldots, x_n$. Deviation of $\tilde{x}_i$ may occur due to measurement errors as well as an absence of strictly linear relationships between variables x and y.

1.3 Statistical inferences

In addition to estimating the parameters of distributions, statistics are also used to infer conclusions regarding data properties [1]. Usually, a statistical inference is the result of testing competitive hypotheses H_i that declare certain data properties. Such a test procedure can be described in the following way:

- formulating the properties (null hypothesis H_0) that need to be confirmed;
- setting the significance level α (usually, $\alpha = 0.05$) for a validation of the null hypothesis;
- calculation of statistics T_n whose value depends on source data and which is used for an inference about the truth of the hypothesis;
- determining a critical region of these statistics where the equality $P(T_n \geq T_{1-\alpha}) = \alpha$ is satisfied;
- confirmation of the null hypothesis truth: if the value of statistics $\hat{T}_n$ is inside (outside) of the critical region $T_{1-\alpha}$, the hypothesis is accepted (rejected).

It is noteworthy to mention that the nature of statistical inference is such that if the hypothesis is accepted this doesn't mean that it has been verified with a given probability. All it means that there is no reason to reject this hypothesis.

The rejection of a hypothesis is accompanied by two kinds errors:

- non-acceptance of a hypothesis even though it is true (the error of the first kind);
- acceptance of a hypothesis even though it is false (the error of the second kind).

If the hypothesis is not accepted then it is possible to predetermine the probably P_α of the error of the first kind. If the hypothesis is accepted, one can determine the probability P_β of the error of the second kind (acceptance of wrong hypothesis) for the alternative hypothesis. The probability $(1 - P_\beta)$ is also called the criterion power.

Consider a few tests that will be referred to in the following text. The first one is the t-test applied to verify equality of the mean values of two normal series $x_{1i}(i = 1, n_1)$ and $x_{2j}(j = 1, n_2)$, that is, to verify the null hypothesis H_0: $\mu_1 = \mu_2$. This test implementation starts with calculations of mean and variance estimates of these realizations:

$$\hat{\mu}_{x_1} = \frac{\sum_{i=1}^{n_1} x_{1i}}{n_1}, \; \hat{\mu}_{x_2} = \frac{\sum_{j=1}^{n_2} x_{2j}}{n_2};$$

$$\hat{D}_{x_1} = \frac{\sum_{i=1}^{n_1} (x_{1i} - \hat{\mu}_{x_1})^2}{n_1 - 1}, \; \hat{D}_{x_2} = \frac{\sum_{j=1}^{n_2} (x_{2j} - \hat{\mu}_{x_2})^2}{n_2 - 1}.$$

If $\hat{\mu}_{x_1} > \hat{\mu}_{x_2}$ the following statistic is calculated:

$$\hat{T}_n = \frac{\hat{\mu}_1 - \hat{\mu}_2}{\sqrt{\hat{D}_{x_1}/n_1 + \hat{D}_{x_2}/n_2}} = \hat{t}_n,$$

where the degrees of freedom n equals

$$n = \frac{\left(\hat{D}_{x_1}/n_1 + \hat{D}_{x_2}/n_2\right)^2}{\frac{\left(\hat{D}_{x_1}/n_1\right)^2}{n_1 - 1} + \frac{\left(\hat{D}_{x_2}/n_2\right)^2}{n_2 - 1}}.$$

If $D_{x_1} = D_{x_2}$, then $n = n_1 + n_2 - 2$.

The obtained value of $\hat{t}_n$ is compared with the tabular $t_{1-\alpha,n}$ corresponding to the given significance level α. The null hypothesis is rejected, if $\hat{t}_n > t_{1-\alpha,n}$.

Another test is the F-test of variance homogeneity. In this case, the null hypothesis declares that two normal variables X_1 and X_2 have the same variance $(H_0\colon D_{x_1} = D_{x_2})$. If $\hat{D}_{x1} > \hat{D}_{x_2}$, then statistic T_n is calculated as:

$$\hat{T}_n = \frac{\hat{D}_{x_1}}{\hat{D}_{x_2}} = \hat{F}_{n_1-1,n_2-1}.$$

The calculated statistic $\hat{F}_{n_1-1,n_2-1}$ is compared with the tabular value $F_{1-\alpha,n_1-1,n_2-1}$ Whenever $\hat{T}_n > F_1 - \alpha, n_1 - 1, n_2 - 1$, the null hypothesis is rejected.

The last test to be considered is used to verify hypotheses about the type of distribution law of random variables. The basis of the test is Pearson's chi-squared test which evaluates the difference between empirical and theoretical frequencies of an event's occurrence. Suppose that the series $x_i(i = 1, n)$ is a result of observing a random variable X. For the null hypothesis, an assumption has to be taken that the empirical data correspond to a certain theoretical distribution. Sometimes, there is sufficient *a priori* knowledge about an observed phenomenon to make such an assumption. In its absence, a histogram of x_i values is constructed. The histogram is a stepped figure consisting of rectangles whose bases are equal to the boundaries of the intervals $x^{(j-1)} \ldots x^{(j)} (j = 1, k)$. The heights of rectangles correspond to the numbers l_j or frequencies $v_j = l_j/n$ of the results falling into intervals. The image of the histogram (Fig. 1.10) may help to make an assumption about the distribution law of values of x_i.

The image of the histogram represented in Fig. 1.10 forms the basis of an assumption of the normal distribution of the variable X for the null hypothesis.

As a next step, the estimates of parameters of the assumed law are calculated using x_i. Let us suppose that the number of unknown parameters is equal m. After that, the range $x_{min} \ldots x_{max}$ is divided into k intervals so that on average each one had $l_j \geq 5$ readouts. The obtained estimates of distribution parameters are used to

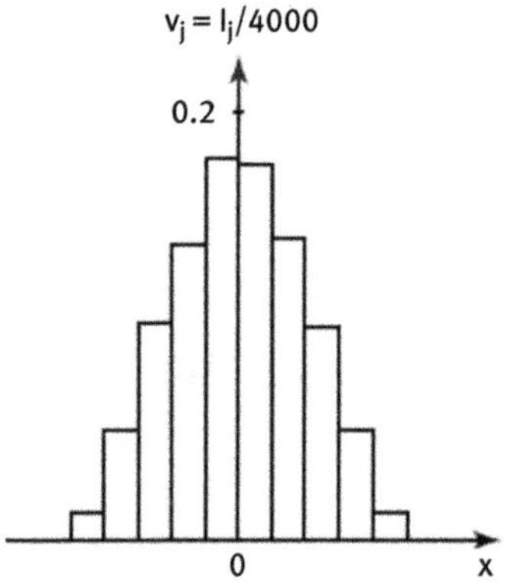

Fig. 1.10: Histogram image.

calculate theoretical values p_j which must be not insignificant. Statistic T_n is formed as follows:

$$\tilde{T}_n = \sum_{j=1}^{k} \frac{(l_j - np_j)^2}{np_j} = n \sum_{j=1}^{k} \frac{(v_j - p_j)^2}{p_j}.$$

According to Pearson's theorem, this statistic has the χ^2 distribution with $(k - m - 1)$ degrees when $n \to \infty$, that is,

$$T_{n\to\infty} = \chi^2_{k-m-1}.$$

As before, the obtained value $\tilde{T}_n$ is compared with the tabular value $\chi^2_{1-\alpha,k-m-1}$ corresponding to the given significance level α. The null hypothesis is rejected, if $\tilde{T}_n > \chi^2_{1-\alpha,k-m-1}$.

2 Metrological model of automated measurements

The approaches and methods presented in this monograph will be illustrated by examples from tests of gas turbine engines (*GTEs*). Therefore, it seems appropriate to provide conceptual data about the *GTE*. The main components of the *GTE* are a multistage compressor (designated as 1 in Fig. 2.1), a combustion chamber (2) and a turbine (3).

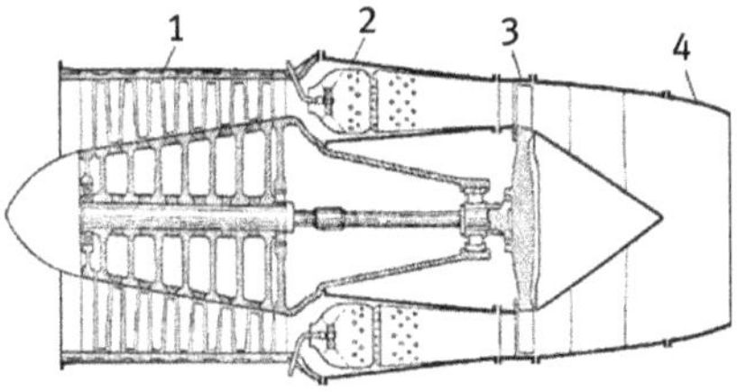

Fig. 2.1: Sketch of the *GTE*.

The compressor is a device that increases the pressure of the air entering it. Combustion chamber adds heat to the compressed air by burning fuel. The turbine converts part of the internal energy of hot, high-pressured gases into mechanical work of the shaft driving the compressor. To this end, the turbine and compressor are rigidly interconnected. If compressor increases the gas pressure, then the turbine decreases the gas pressure. The remaining energy of gases can be used in different ways.

For example, in aviation applications the gases can be directed into a nozzle (designated as 4 in Fig. 2.1); in this narrowing channel the potential energy of the gas is converted into kinetic energy. As in the turbine, gas expansion occurs in the nozzle. The emerging jet stream creates traction (thrust), which moves the plane forward. Such an engine is called a turbojet.

In another type of engine (turboprop), the second turbine (called a free one) is installed to utilize the remaining energy of gases. The shaft of this turbine is connected to the propeller through a reduction gearbox. The final expansion of gases exhausting from the free turbine occurs in the nozzle where their pressure drops to atmospheric levels. The main thrust provided by turboprops is created by the air flow accelerated by the propeller. The thrust provided by the nozzle converted to equivalent power does not exceed 10% of the engine's thrust power.

An example of the *AMIS* utilized in the *GTE* test facility is presented in Fig. 2.2.

The *AMIS* is built around a computer that communicates with the test object, equipment and technological systems by means of a data input/output device (designated as 1 in Fig. 2.2). Its task is to accept measurement signals and to send control commands.

https://doi.org/10.1515/9783110666670-003

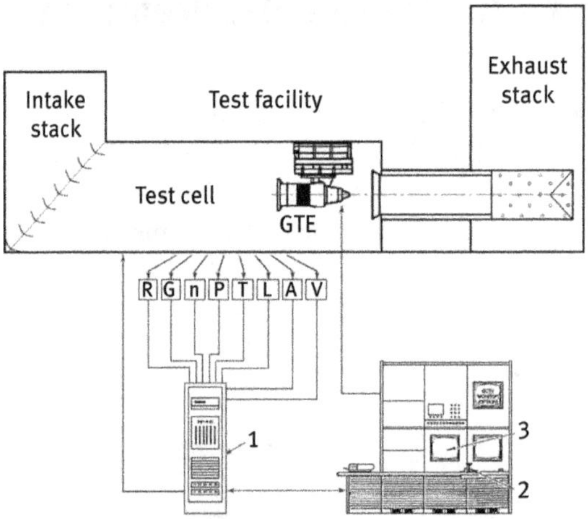

Fig. 2.2: Sample scheme of the *AMIS*.
(1 – data input/output device; 2 – engine throttle lever; 3 – operator screen)

Accepted measurement signals correspond to the following physical values:
- Engine thrust (*R*);
- Fluids (fuel, oil, hydrolic liquids) flow (*G*);
- Shaft rotation speed (*n*);
- Pressures of air, gases and fluids (*P*);
- Temperatures of air, gases and fluids (*T*);
- Linear (*L*) and angular (*A*) displacements;
- Vibrations (*V*), etc.

Parameters (pressures and temperatures) of the air and gases are measured at cross-sections corresponding to inlet and outlet of the *GTE* as well as to interface points of the above engine components.

Besides acquisition of measurement signals, the data input/output device supplies commands to the Engine Control Unit (*ECU*) as well as to *PLC*s and actuators of technological systems and test equipment. Thus, the required test conditions and operating modes of the *GTE* are provided.

The control panel with the engine throttle lever and operator displays (designated as 2 & 3 in Fig. 2.2) supports the interactive work of test personnel (control and monitoring of the test object and facility systems). Appropriate peripheral equipment of the *AMIS* provides hard copies of test results (test reports) as well as storage of test results in archives.

As it was mentioned in the Introduction section, the *AMIS* creates a unified information environment for formatting measurement results, their processing and storage as well as and other data manipulations. Therefore, the metrological structure of the system may be presented as shown in Fig. 2.3 [2].

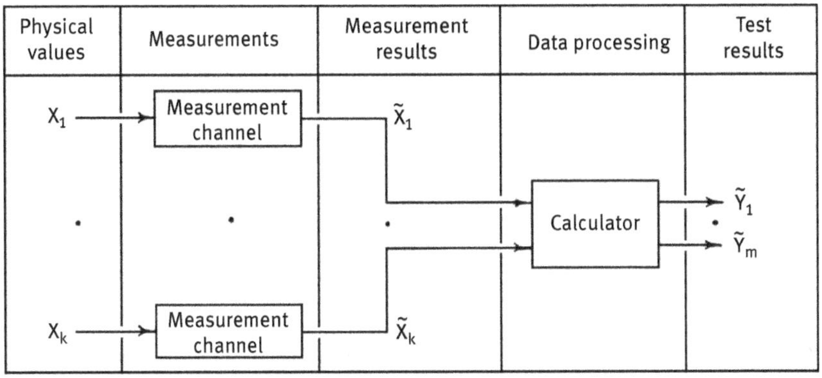

Fig. 2.3: Metrological structure of the *AMIS*.

A measurement channel is a functionally united component of the system providing estimates of measured values in digital form. The outputs of all measurement channels are the result of direct measurements $\widetilde{\boldsymbol{X}}^{\mathrm{T}} = \left(\widetilde{X}_1, \ldots, \widetilde{X}_k\right)$.

Subsequent data processing realizes calculations of parameters determining operational conditions as well as engine parameters and performance. This procedure is executed by a processor that works in accordance with software written in some programming language or commercial software products (*COTS*) with abilities for data processing. Therefore, the test results $\widetilde{\boldsymbol{Y}}^{\mathrm{T}} = (\widetilde{Y}_1, \ldots, \widetilde{Y}_m)$ are defined as results of indirect and aggregate measurements.

2.1 Measurement channel

From the technical standpoint, the measurement channel is a kit of interconnected probes/sensors, transduces, convertors, acquisition devices as well as processor/software that implements computational functions. A typical diagram of the *MC* may be represented by a channel measuring total pressure P^* of an air stream (Fig. 2.4).

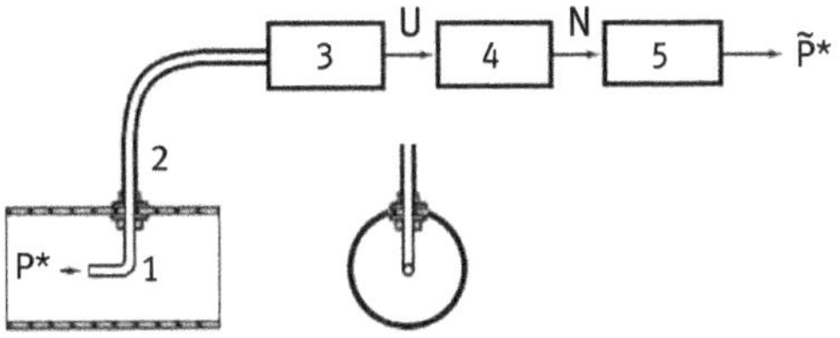

Fig. 2.4: Conditional scheme of a measurement channel.

Probe 1 localizes the measurement cross-section as well as a measurement point in the cross-section plane.

Air pipeline 2 "supplies" a value P^* to a sensor of the pressure transducer 3.

The pressure value affects a sensor (membrane) of the transducer. An array of silicon piezo resisters combined in a bridge circuit is mounted on the membrane. Its deformation corresponding to the value of P^* yields a change of values of piezo resisters. As a result, the value of P^* transforms to a value of a direct current voltage U.

To be able to use signal $U(t)$ in the *AMIS*, its values must be sampled at moments in time (t_i) due to the discrete nature of computers. The conversion of a continuous signal $U(t)$ into a sequence of numbers N is carried out by the Analog-Digital Converter (*ADC*). The *ADC* (designated as 4 in Fig. 2.4) realizes two consistent operations:

- discretization of signal $U(t)$ in time which results in readouts $u_i = U(t_i)$;
- digitizing of values of u_i in numbers N_i.

From a mathematical standpoint, digitization deals with value u_i round off according to specified rules. Let values of u_i be inside a range $0 \ldots U_{max}$. Within this range, one can fix M values $u^{(1)}, u^{(2)}, \ldots, u^{(M)}$ called quantization levels. The digitization procedure consists of pairing the value of u_i with one of the quantization levels. A commonly used practice is an identification of the value of u_i with nearest low quantization level u^j. As it is known, for digital techniques the binary system is usually used, so the number of quantization levels corresponds to the base power 2, that is, $M = 2^m - 1$. As a result, the *ADC* output N is an m-bit binary code (word).

The round-off operation means that the digitized value of u_i will have an error. When values of u_i are identified with the nearest low level, this quantization error $\varepsilon_q = u_i - u^{(j)}$ will have a value within the $\Delta u = u^{(j+1)} - u^{(j)}$ range. Over a small interval Δu (actually, $M > 1000$) the probability density $f(\varepsilon_q)$ may be considered constant. It means that all values of ε_q have a virtually equal probability, that is, $f(\varepsilon_q)$ corresponds to a uniform distribution (Fig. 2.5).

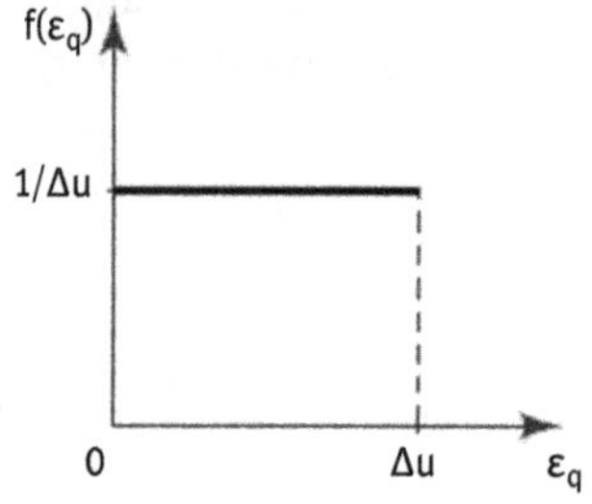

Fig. 2.5: The *PDF* of quantization error.

The assumption of the uniform distribution of quantization error ε_q allows to determine its mathematical expectation as:

$$\mu_q = \int_0^{\Delta u} \varepsilon_q f(\varepsilon_q) d\varepsilon_q = \Delta u/2.$$

The value of Δu is always known *a priori*, therefore, a systematic error $\mu_q = \Delta u/2$ can be assumed in subsequent processing.

Because of the uniform distribution of quantization error, its variance will be equal to

$$D_q = \int_0^{\Delta u} (\varepsilon_q - \Delta u/2)^2 f(\varepsilon_q) d\varepsilon_q = (\Delta u)^2/12.$$

A modern *ADC* has $m \geq 16$ $(M \geq 65535)$, and a corresponding value $\Delta u/U_{max} \leq$ 0.000015, that is, 0.0015% of a full scale (*FS*). It should be noted that tests of complex manufactured objects are characterized by measurement error of $\geq 0.1\%$. So, for a practical purpose the digitizing effect may be ignored.

A readout N_i is acquired via a digital interface in the computer (designated as 5 in Fig. 2.4). The latter using a calibration curve $P^* = g(N)$ transforms N_i into a measurement result $\tilde{P}^*$ represented as:

$$\tilde{P}^* = P^* + \Xi_{P^*},$$

where Ξ_{P^*} is a measurement error. In common cases, a measurement channel model is described as:

$$\tilde{X}_i = X_i + \Xi_i.$$

2.2 Processing of measurement results

In the *AMIS*, measurement results are processed to evaluate parameters determined operational conditions of a test object as well as its performance. Technically, this procedure is executed by a computer processor working in accordance with software that realizes appropriate processing algorithms.

Besides calculations of results of indirect measurements, data processing may include corrections of direct measurement results. For example, if a working fluid of a test object (gas turbine, turbojet, internal-combustion engine etc.) is inlet air, its thermodynamic properties depend on the atmospheric conditions (pressure, temperature, humidity). To compare results of tests for different calendar days, measurement results are corrected to International Standards Organization (*ISO*) reference conditions. They are sea level, static, standard day atmosphere conditions characterized by values of pressure, temperature, and humidity equal 101325 N/m^2, 288.15 K, 60%, accordingly. For instance, the following formulas are used to correct results of measurements of temperature ($\tilde{T}$), pressure ($\tilde{P}$), velocity ($\tilde{V}$), mass flow ($\tilde{G}$) and power ($\tilde{R}$) [3]:

$$\tilde{T}_{cor} = \tilde{T} \cdot \frac{288.15}{\tilde{T}_{inlet}};$$

$$\tilde{P}_{cor} = \tilde{P} \cdot \frac{101325}{\tilde{P}_{inlet}};$$

$$\tilde{V}_{cor} = \tilde{V} \cdot \sqrt{\frac{288.15}{\tilde{T}_{inlet}}};$$

$$\tilde{G}_{cor} = \tilde{G} \cdot \frac{101325}{\tilde{P}_{inlet}} \cdot \sqrt{\frac{\tilde{T}_{inlet}}{288.15}};$$

$$\tilde{R}_{cor} = \tilde{R} \cdot \frac{101325}{\tilde{P}_{inlet}} \cdot \sqrt{\frac{\tilde{T}_{inlet}}{288.15}}.$$

Therefore, a result of indirect or corrected measurements may be represented as a function of direct measurement results:

$$\tilde{Y}_i = g_i(\tilde{X}_1, \ldots, \tilde{X}_k), \quad i = 1, m.$$

Linearization of this function at a point of actual (true) values $\boldsymbol{X}^{\mathrm{T}} = (X_1, \ldots, X_k)$ yields the following expression:

$$\tilde{Y}_i = g_i(X_1, \ldots, X_k) + \sum_{j=1}^{k} a_{ij} \cdot (\tilde{X}_j - X_j),$$

where $a_{ij} = \partial Y_i / \partial X_j|_X$. Taking into account that $Y_i = g_i(X_1, \ldots, X_k)$ and $\tilde{X}_j = X_j + \Xi_j$, a model for an error E_i of a test result $\tilde{Y}_i$ can be represented in the following form:

$$\mathrm{E}_i = \tilde{Y}_i - g_i(X_1, \ldots, X_k) = \tilde{Y}_i - Y_i = \sum_{j=1}^{k} a_{ij} \cdot \Xi_j.$$

Accordingly, the errors of all test results can be described in a matrix form as:

$$\mathbf{E} = \boldsymbol{A}_c \, \Xi,$$

where $\mathbf{E}$ and Ξ are vectors of errors of measurement and test results. The matrix $\boldsymbol{A}_c$ means a mathematical operator of calculations:

$$\boldsymbol{A}_c = \begin{pmatrix} a_{11} & . & a_{1k} \\ . & . & . \\ a_{m1} & . & a_{mk} \end{pmatrix}.$$

In other words, the metrological model of the processing procedure is reduced to just error propagation.

3 Steady-state performance

Metrological models presented in the previous chapter may be implemented for analysis of test result accuracy. As it was demonstrated by Fig. 2.3, test results (estimates of object parameters/characteristics) are an “output” of the *AMIS*. The presented metrological structure of the *AMIS* allows to realize an accuracy analysis in two successive steps.

The first one is a field procedure for estimating metrological characteristics of measurement channels. It is realized without a reference to a specific test object or a test program. Knowing metrological characteristics of measurement channels allow to evaluate errors Ξ of measurement results $\tilde{\boldsymbol{X}}^{\mathrm{T}} = (\tilde{X}_1, \ldots, \tilde{X}_k)$.

The second step of metrological studies is related to estimating accuracies of test results $\tilde{\boldsymbol{Y}}^{\mathrm{T}} = (\tilde{Y}_1, \ldots, \tilde{Y}_m)$. This procedure has to utilize the known measurement errors Ξ and operator $\boldsymbol{A}_c$. As elements a_{ij} of the matrix $\boldsymbol{A}_c$ have to be estimated for determined value of $\tilde{\boldsymbol{Y}}$, test result accuracy may be done only for tested modes of a specific test object.

Diversity of types of measurement channels and variety of processing algorithms require using a universal research method for above mentioned metrology studies. Such a method that is suitable to the random nature of errors is statistical tests, the essence of which consists of forming realizations of random variables to determine characteristics of their distributions.

3.1 Metrology of measurement channel

From the metrological model of the *MC* it follows that errors Ξ can be determined from the known values of $\tilde{X}$ and X:

$$\Xi = \tilde{X} - X.$$

Since errors are of a random nature, it is necessary to have a sequence of values $\xi_1, \ldots, \xi_n$ to accurately describe them. This sequence can be obtained during statistical tests of channels under operating (field) conditions (Fig. 3.1).

During these field tests, a reference value X is reproduced at the input of a measurement channel. As this value is a standard, a distribution of random value $\tilde{X}$ corresponds to distribution of errors Ξ in a working environment. Therefore, evaluating metrological characteristics of a measurement channel is reduced to an estimation of parameters of the *PDF* $f(\xi)$.

https://doi.org/10.1515/9783110666670-004

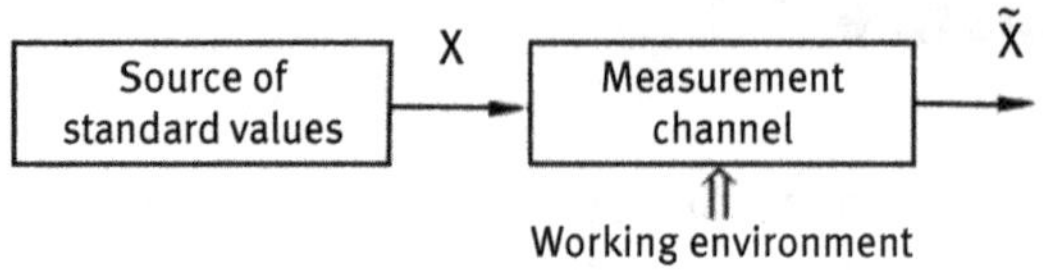

Fig. 3.1: Statistical tests scheme.

3.1.1 Statistical tests of measurement channel

First of all, a requirement for accuracy of reference values has to be formulated. Actually, a reference value X is reproduced with an error $\Theta \neq 0$ that modifies a channel model to the following view:

$$\tilde{X} = (X + \Theta) + \Xi = X + \Sigma,$$

that is, an actual error of a measurement result will be the sum $\Sigma = \Theta + \Xi$. There is a rule which establishes a negligible influence of a random variable Θ compared to another variable Ξ: if $\sigma_\xi \geq 3\sigma_\theta$, the effect of Θ is minimized. This rule is based on the following considerations. In the case where $\sigma_\xi = 3\sigma_\theta$, the *rms* error of the sum Σ equals

$$\sigma_\Sigma = (\sigma_\theta^2 + \sigma_\xi^2)^{1/2} = (\sigma_\theta^2 + 9\sigma_\theta^2)^{1/2} \approx 3.16\sigma_\theta.$$

Hence, ignoring σ_θ leads to the error $\delta_\Sigma = -0.16/3.16 = -0.05$ (5%) in the value of σ_Σ.

In the case of a normal distribution of error Ξ, the *MLE* of its *rms* deviation is determined by the following expression:

$$\hat{\sigma}_\xi = \sqrt{\frac{\sum_{i=1}^{n} (\xi_i - \mu_\xi)^2}{n}} = \sqrt{\frac{\sum_{i=1}^{n} D_\xi z_i^2}{n}} = \sigma_\xi \times \sqrt{\frac{\chi_n^2}{n}} = q\sigma_\xi,$$

where $q = \sqrt{\chi_n^2/n}$.

As it was mentioned in section 1.2.1, a variable $\zeta = \sqrt{2\chi_n^2}$ may be considered normal with parameters:

$$\mu_{\chi_n^2} = \sqrt{2n - 1};$$

$$D_{\chi_n^2} = 1,$$

when $n > 30$. As a result, corresponding parameters of the variable $q = \zeta/\sqrt{2n}$ are:

$$\mu_q = \sqrt{\frac{2n-1}{2n}} \approx 1;$$

$$\sigma_q = 1/\sqrt{2n}.$$

As q has a normal distribution, its confidence $(P = 0.95)$ interval Δ_q will have the following lower ("$\leq$") and upper ("$\geq$") bounds:

$$\Delta_{q\leq} = 1 - \frac{1.96}{\sqrt{2n}}; \qquad \Delta_{q\leq} = 1 + \frac{1.96}{\sqrt{2n}}.$$

Hence, the error δ_ξ of the $\hat{\sigma}_\xi$ estimates can be written as:

$$\delta_\xi = \frac{1.96}{\sqrt{2n}}.$$

Finally, the size n of a statistical series used for the estimation of σ_ξ can be represented as a function of δ_ξ:

$$n = \frac{1.92}{\delta_\xi^2}.$$

Earlier, it was demonstrated that for the case $\sigma_\xi = 3\sigma_\theta$, ignoring the σ_θ value leads to error $\delta_\Sigma = -0.05$ in estimates of σ_Σ. A suitable *rms* error $(\delta = 0.05)$ of $\hat{\sigma}_\xi$ can be achieved for a realization with size $n > 768$. During tests of measurement channels, the number n is usually on the order of a hundred. In this case, a statistical variation (14%) of σ_ξ will exceed the effect of ignoring the value σ_θ by almost ~3 times. This fact is the basis for the requirement to reproduce reference values with an accuracy of at least three times greater than the accuracy of the tested channel.

Now, it is time to describe the channel test procedure itself. During statistical tests the reference values $x_i (i = \overline{1, l})$ vary from X_{min} to X_{max} (run up) and back from X_{max} to X_{min} (run down). Such an approach makes it possible to reveal a systematic change in the measurement results due to the possible phenomenon of hysteresis in sensor materials. For example, in the pressure measurement channel described in Section 2.1, the sensitive element is a membrane deformed under pressure. Internal friction of membrane material may cause the energy dissipation which leads to the appearance of an elastic hysteresis an example of which is represented in Fig. 3.2.

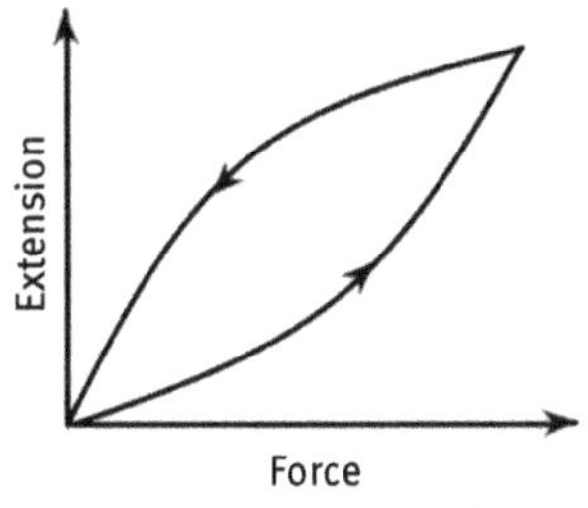

Fig. 3.2: Conditional presentation of elastic hysteresis.

For reference value x_i, reproduced during the run up and down, readouts N_{ij} $(j=1,n)$ at the output of the *ADC* are formed. This data is registered in computer memory. Therefore, the statistical test of a measurement channel will result in formation of the $2l$ series of binary numbers $N_{ij\rightarrow}$ and $N_{ij\leftarrow}$. The indexes "→" and "←" correspond to movement from a small value of x_i to a large one and back, accordingly.

Processing results of a statistical test begins with a determination of a calibration curve $X=g(N)$. It should be noted that another name for statistical tests is channel calibration, and the results are called calibration data.

To determine the calibration curve, the grouping centers of numbers $N_{ij\rightarrow}$ and $N_{ij\leftarrow}$ are calculated as:

$$\hat{N}_i = \frac{\sum_{j=1}^{n}(N_{ij\rightarrow}+N_{ij\leftarrow})}{2n}.$$

Thus, pairs x_i and $\hat{N}_i$ describing the points of a calibration curve are formed. The analytical (formal) representation of the calibration curve can be obtained, for instance, by the least squares method.

At this stage it is advisable to determine the size of statistical series, that is, the value n. It was demonstrated in section 1.2.3 that a variance of estimates of a mathematical expectation equals to the variance of a time series divided by its size. Therefore, the estimate $\hat{x}_i = g(\hat{N}_i)$ will have the *rms* error $\sigma_{\hat{x}_i} = \sigma_\xi/\sqrt{2n}$.

As it was stated in section 1.2.4, using the *LSM* for approximation of a certain function requires reliable knowledge of its arguments. In other words, argument values have to have zero or negligible errors. This condition can be formulated as a previously mentioned rule for the ratio of reference and measurements error $-\sigma_\xi \geq 3\sigma_\theta$. Since the choice of this ratio is perceived as subjective, we take its value of 3 resulting in the following expression for σ_θ:

$$\sigma_\theta = 3\sigma_{\hat{x}_i} = \frac{3\sigma_\xi}{\sqrt{2n}} \geq \frac{9\sigma_\theta}{\sqrt{2n}}.$$

The modification of this inequality yields a realization size $n \geq 41$. Long-term practice of the author has shown that the value $n=100$ provides sufficient stability of statistical estimates and acceptable time spent for the calibration procedure as well as processing of calibration data.

Presence of pairs of values x_i and $\hat{N}_i$ can be used to estimate factors of a polynomial function which will approximate the calibration curve:

$$X = a_0 + a_1 N + \;\dots\; + a_p N^p.$$

Usually, the polynomial $p \leq 3$, and its value is found as a result of its successive increase from value $l=1$ until obtaining the result

$$\max_i \left|\tilde{X}_i - X_i\right| < \Delta_\theta.$$

Here Δ_θ is a confidence interval for $P = 0.95$. In the case of a normal distribution of Θ, $\Delta_\theta = 1.96\sigma_\theta$.

As far as the choice of a number l and reference values x_i is concerned, it is somewhat arbitrary. However, the following considerations may be the basis of such a choice. The form of $X = g(N)$ is determined mainly by the calibration curve of the primary sensor used for the measurement channel. Therefore, the choice of the values of l and x_i can be made by taking into account the shape of a sensor calibration curve. Usually, $l \geq 10$, and the placement of x_i should give an idea of the features of the curve, that is, reference values should be located with smaller intervals in areas with significant curvature. One point should be in the middle of the measurement range in order to detect possible hysteresis.

Another form of analytical description of a calibration curve is its piecewise linear approximation (Fig. 3.3).

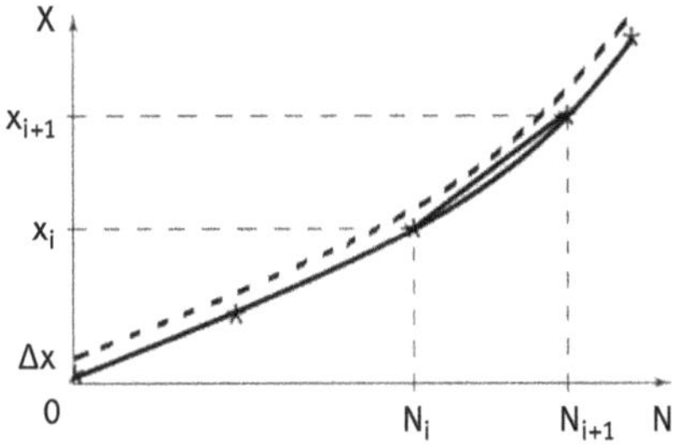

Fig. 3.3: Piecewise linear approximation.

In this case, transforming result N_j into a measured value starts with searching an interval $N_i \leq N_j < N_{i+1}$. Subsequently, the value $\tilde{x}_j$ is calculated as:

$$\tilde{x}_j = x_i + \frac{x_{i+1} - x_i}{N_{i+1} - N_i} \cdot (N_j - N_i).$$

Comparing with a polynomial function which requires the storage of factors $\hat{a}_0, \ldots, \hat{a}_p$, the piecewise linearization requires a storage of calibration data in a form of l pairs of x_i and N_i. The advantage of this representation of the calibration curve is in the possibility of providing a small approximation error Δ_x by increasing the number of control points.

After obtaining a formal description of the calibration curve, binary numbers $N_{ij\rightarrow}$ and $N_{ij\leftarrow}$ are converted to the series $\tilde{x}_{ij\rightarrow}$ and $\tilde{x}_{ij\leftarrow}$ $(i = 1, l; j = 1, n)$. These data and standard values x_i are stored in the archives for further processing in order to obtain estimates of the metrological characteristics of the channel.

The multichannel *AMIS* implies the repetition of calibration procedures, which motivates automating the channel calibration. This allows to reduce the staff involved for these tasks, to reduce the time required to carry out the procedure and to increase the reliability of the data. Therefore, the system has a dedicated mode of operation that enables *AMIS* to use some of its resources for an automated calibration procedure, that is, the reproduction of reference values, measurements, data collection and storage.

3.1.2 Estimation of metrological characteristics

As it's stated earlier, metrological performance of channels is related to parameters of measurement error distributions. Therefore, the purpose of processing calibration data is estimation of $f(\xi)$ parameters utilizing the results of the calibration procedure.

Processing at each ith point of the measurement scale begins with calculating the grouping centers of readouts during both run up and down:

$$\hat{x}_{i\rightarrow} = \frac{\sum_{j=1}^{n} \tilde{x}_{ij\rightarrow}}{n}$$

$$\hat{x}_{i\leftarrow} = \frac{\sum_{j=1}^{n} \tilde{x}_{ij\leftarrow}}{n}.$$

Calibration data corresponding to a reference value x_i and its processing results are illustrated by Fig. 3.4.

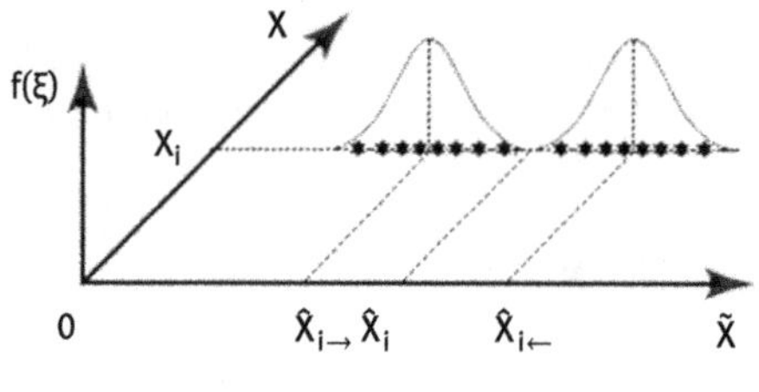

Fig. 3.4: Illustration of calibration data.

The next action is estimating the channel variation $\hat{v}_i$:

$$\hat{v}_i = \hat{x}_{i\leftarrow} - \hat{x}_{i\rightarrow}.$$

Following that action, the values of errors ξ_{ij} are calculated:

$$\xi_{ij\rightarrow} = \tilde{x}_{ij\rightarrow} - \hat{x}_{i\rightarrow};$$

$$\xi_{ij\leftarrow} = \tilde{x}_{ij\leftarrow} - \hat{x}_{i\leftarrow},$$

which allows to estimate their *rms* deviations:

$$\hat{\sigma}_{\xi_{i\rightarrow}} = \sqrt{\frac{\sum_{j=1}^{n} \xi_{ij\rightarrow}^{2}}{n-1}};$$

$$\hat{\sigma}_{\xi_{i\leftarrow}} = \sqrt{\frac{\sum_{j=1}^{n} \xi_{ij\leftarrow}^{2}}{n-1}}.$$

This procedure must be implemented on l points of a measurement scale.

For the point $i = 1$, an additional action is performed – the homogeneity test of variances corresponded to run up and down data. Recall that from point $i = 1$ the calibration procedure starts and ends at that point. The F-test is made necessary by the need to verify that the conditions of the calibration procedure did not change or that their changes did not have a significant effect on the MC characteristics. If $\hat{\sigma}_{\xi_{i\leftarrow}} > \hat{\sigma}_{\xi_{i\rightarrow}}$, then statistic T_n is calculated as:

$$\hat{T}_n = \frac{\hat{\sigma}_{\xi_{i\leftarrow}}}{\hat{\sigma}_{\xi_{i\rightarrow}}} = \hat{F}_{n-1,n-1}.$$

The obtained value of $\hat{F}_{n-1,n-1}$ must be compare with the tabular $F_{0.95,n-1,n-1}$. If the test demonstrates that the null hypothesis H_0: $\sigma_{\xi_{1\rightarrow}} = \sigma_{\xi_{1\leftarrow}}$ is rejected, then the calibration procedure is analyzed for its qualitative performance and for detecting causes that could lead to a trend in the values of error variances.

In the case of a positive conclusion for the null hypothesis (variances are considered equal), the above listed calculations of variations and variances are performed for the next points $(i = 2, l)$ of the scale.

After processing the calibration data at all points of the measurement scale, errors $\xi_{ij\rightarrow}$ and $\xi_{ij\leftarrow}$ may be combined in one set. Respectively, the channel *rms* error $\hat{\sigma}_{ch}$ may be estimated as:

$$\hat{\sigma}_{ch} = \sqrt{\frac{\sum_{i=1}^{l} \sum_{j=1}^{n} \left(\xi_{ij\rightarrow}^{2} + \xi_{ij\leftarrow}^{2}\right)}{2l(n-1)}} = \sqrt{\frac{\sum_{i=1}^{l} \left(\hat{\sigma}_{\xi_{i\rightarrow}} + \hat{\sigma}_{\xi_{i\leftarrow}}\right)}{2l}}.$$

Applying the pessimistic approach, an upper bound ($P = 0.95$) of the $\hat{\sigma}_{ch}$ may be estimated in the following way:

$$\hat{\sigma}_{\tilde{X}} = \left(1 + \frac{1.96}{\sqrt{2l(n-1)}}\right) \hat{\sigma}_{ch}.$$

Subsequently, the greatest value of the channel variation is determined:

$$\hat{v}_{ch} = \max_{i=1,l} \hat{v}_i.$$

The value of $\hat{v}_{ch}$ must be tested for its equality to zero, that is, for its insignificance. In other words, a test for the null hypothesis H_0: $\hat{x}_{i\leftarrow} = \hat{x}_{i\rightarrow}$ has to confirm that the calculated $\hat{v}_{ch} = \hat{x}_{i\leftarrow} - \hat{x}_{i\rightarrow} \neq 0$ only due to statistical variability of measurement results.

Since grouping centers $\hat{x}_{i\leftarrow}$ and $\hat{x}_{i\rightarrow}$ are sums of $n > 40$ values of variable x_i, the central limit theorem allows to speculate about the Gauss law for the estimates of $\hat{v}_{ch}$. Therefore, the t-test can be utilized to verify the null hypothesis. Recall that *rms* deviations of both grouping centers $\hat{x}_{i\rightarrow}$ and $\hat{x}_{i\leftarrow}$ are:

$$\hat{\sigma}_{\hat{x}_{i\leftarrow}} = \hat{\sigma}_{\hat{x}_{i\rightarrow}} = \hat{\sigma}_{ch}/\sqrt{n}.$$

Therefore, the calculated statistic T_n will have the following form:

$$\hat{T}_n = \frac{\hat{x}_{i\leftarrow} - \hat{x}_{i\rightarrow}}{\sqrt{\frac{\hat{D}_{\hat{x}_{i\leftarrow}}}{n} + \frac{\hat{D}_{\hat{x}_{i\rightarrow}}}{n}}} =$$

$$\frac{\hat{v}_{ch}}{\hat{\sigma}_{ch}\sqrt{2/n}} = \hat{t}_{2n}.$$

The value $\hat{t}_{2n}$ must be compare with the tabular $t_{0.95,2n}$. Depending on the results of the t-test, two options are possible for calculating the accuracy indicator of the calibrated channel.

The first option corresponds to the situation when there is no reason to reject the null hypothesis H_0: $\hat{x}_{i\leftarrow} = \hat{x}_{i\rightarrow}$, that is, the value of $\hat{v}_{ch} = 0$. In this case, the distribution law $f(\xi)$ can be identified using the Pearson's chi-squared test in order to determine an accuracy indicator (the confidence interval $\Delta_{\tilde{X}}$ for $P = 0.95$). If a distribution low is different from the typical ones (normal, triangular, uniform etc.), a histogram of scattering $\xi_{ij\rightarrow}$ and $\xi_{ij\leftarrow}$ can be constructed in order to get an empirical estimate of $\Delta_{\tilde{X}}$.

If the null hypothesis fails then there is variation between measurements corresponding to run up and run down. In this case, the following approach can be used. For specific measurement during an object's tests, the value of v_i is uncertain due to changing modes (run up and down) of the object as well as carrying out measurements at different points of a measurement scale. In other words, the test procedure randomizes values of v_{ch}. Therefore, for a particular measurement, the variation can be considered a random variable with a uniform distribution law at the range $\mp v_{ch}/2$.

Because the variation estimate has a uniform distribution, it may be characterized by the *rms* deviation:

$$\hat{\sigma}_v = \sqrt{\frac{\hat{v}_{ch}^{\,2}}{12}}.$$

Finally, the *rms* error of the measurement results $\tilde{X}$ accompanied by variations can be evaluated as:

$$\hat{\sigma}_{\tilde{X}} = \sqrt{\hat{\sigma}^2_{ch} + \hat{\sigma}^2_{v}}.$$

As in the previous case, a histogram of scattering values $(\tilde{x}_{ij\rightarrow} - \hat{x}_i)$ and $(\tilde{x}_{ij\leftarrow} - \hat{x}_i)$ can be constructed using the parameter $\hat{\sigma}_{\tilde{X}}$. As a result, the indicator $\Delta_{\tilde{X}}$ of channel accuracy can be found in an empirical fashion.

The *rms* (standard) deviation $\hat{\sigma}_{\tilde{X}}$ and confidence interval $\Delta_{\tilde{X}}$ characterize the instrumental error of a single measurement. Their values are determined by:

- the resolution of a primary transducer of the measured physical value,
- characteristics of channel components and their stability,
- methods and quality of calibration procedure.

3.2 Analysis of test result accuracy

As it was mentioned earlier, a given mode of the test object is characterized by its performance estimated by processing measurement results $\tilde{\boldsymbol{X}}^{\mathrm{T}}(\tilde{X}_1, \ldots, \tilde{X}_k)$:

$$\tilde{\boldsymbol{Y}} = \boldsymbol{A}_c\tilde{\boldsymbol{X}},$$

where $\tilde{\boldsymbol{Y}}^{\mathrm{T}} = (\tilde{Y}_1, \ldots, \tilde{Y}_m)$ are test results. The operator of calculations $\boldsymbol{A}_c$ is a matrix

$$\boldsymbol{A}_c = \begin{pmatrix} a_{11} & . & a_{1k} \\ . & . & . \\ a_{m1} & . & a_{mk} \end{pmatrix}$$

with elements $a_{ij} = \frac{\partial Y_i}{\partial X_j}$ whose numerical value is calculated by substitution of measurement results $(\tilde{X}_{1l}, \ldots, \tilde{X}_{kl})$ obtained for the *lth* object mode.

The test results are characterized by the covariance matrix $\boldsymbol{\Gamma}_{\tilde{Y}}$:

$$\boldsymbol{\Gamma}_{\tilde{Y}} = \boldsymbol{A}_c\boldsymbol{\Gamma}_{\tilde{X}}\boldsymbol{A}_c^{\mathrm{T}},$$

where $\boldsymbol{\Gamma}_{\tilde{X}}$ is the covariance matrix of measurement results. The latter is compiled using the known metrological characteristics of the measurement channels:

$$\boldsymbol{\Gamma}_{\tilde{X}} = \begin{pmatrix} \hat{\sigma}^2_{\tilde{X}_1} & . & 0 \\ . & . & . \\ 0 & . & \hat{\sigma}^2_{\tilde{X}_k} \end{pmatrix},$$

It is a diagonal matrix due to independence of results of direct measurements $\tilde{X}_{1l}, \ldots, \tilde{X}_{kl}$.

Hence, the knowledge of the matrices $\mathbf{\Gamma}_{\tilde{X}}$ and $\boldsymbol{A}_c$ at a given mode of object operation allows to calculate the covariance matrix of test results. The main disadvantage of this approach is associated with the multi-factor nature of the task. In addition to the values of k and m being the task factors, it is need to take into account various dependencies and constants obtained experimentally with their own uncertainties. In addition, the calculation of the object characteristics is carried out taking into account its geometrical sizes as well as features of the object layout on the test stand. Such complexity of the multi-factor nature of the task makes it difficult to evaluate the accuracy of test results through mathematical calculations.

3.2.1 Simulation of measurement errors

The logical continuation of the metrological studies of the *MC*s described in section 3.1 is the use of the method of statistical tests. Since the metrological model of data processing is just a propagation of measurement errors, the latter have to be reproduced during statistical tests. This can be implemented by simulating measurement errors with random numbers. In other words, statistical modeling of the measurement phenomenon by Monte Carlo method [4] should be carried out. To utilize this method, one can use the random number generator which is available in the processor of modern computers. For example, in the *Intel* chips, such a generator known as *RdRand* supplies random numbers at the request of software. As a rule, these numbers have the uniform distribution. By virtue of the central limit theorem, their sum gives a normal distribution. For practical purposes, it is enough to sum 12 numbers with a uniform distribution. The numbers thus formed are called pseudo-random, since they are only approximations of true random numbers, at least due to the fact that they have a period of repetition of their values. A generator supplying such numbers is called the pseudo-random number generator (*PRNG*).

The essence of statistical modeling is as follows. The operator $\boldsymbol{A}_c$ can be considered a model of a certain hypothetical system with determined numbers of inputs and outputs. In addition to the results of measurements, the inputs of the system may be, for example, geometric sizes, constants, dependencies etc. The outputs of the system are estimated characteristics of the test object. Statistical simulation of the errors of the system inputs allows to obtain the realizations of its outputs. In principle, both instrumental and methodical errors can be simulated. To do this, as it was stated earlier, the *PRNG* is used to form a set of random numbers Ξ. Their distribution type and parameters correspond to the errors ξ_i of the measurement result $\tilde{x}_i$. The set of numbers Ξ are added to the values of the system inputs $\tilde{\boldsymbol{X}}$ corresponding to the given mode of the test object. The obtained values $(\tilde{\boldsymbol{X}}+\Xi_j)$ are processed by the operator $\boldsymbol{A}_c$ providing the corresponding values $\tilde{\boldsymbol{Y}}_j$ (Fig. 3.5).

Repeating this procedure n times $(j=\overline{1,n})$ allows to obtain a series of outputs $\tilde{y}_{i1},\dots,\tilde{y}_{in}$. Their statistical properties will characterize the accuracy of test results.

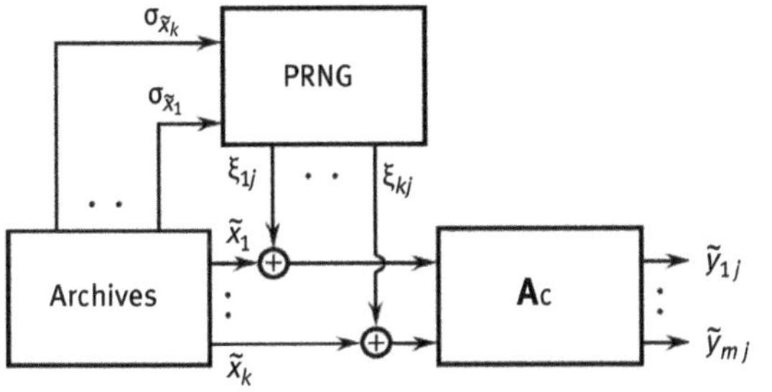

Fig. 3.5: Statistical simulation of measurement errors.

The implementation of the Monte Carlo method is related to the consideration of the choice of size n for statistical series as well as the method of modeling the errors of the measurement results. As far as the number n is concerned, in section 3.1.1 a confidence interval $(P = 0.95)$ was determined for errors of σ_ξ estimates in the following form:

$$\Delta_{\hat{\sigma}_\xi} = \frac{1.96}{\sqrt{2n}}.$$

This expression allows to determine that $n = 800$ provides an uncertainty for estimates of σ_ξ equals $\Delta_{\hat{\sigma}_\xi} < 0.05\,(5\%)$ which is quite acceptable for engineering applications.

An approach for modeling of measurement errors may be as follows. There are two possibilities for describing the accuracy of measurement results. The first corresponds to the case set forth in section 3.1.1 when metrological characteristics result from statistical tests of measurement channels. In this case, errors of the *MC*s may be simulated by pseudo-random numbers with the Gauss distribution law and *rms* deviation $\hat{\sigma}_{ch}$. With the same deviation, normal variables have the largest uncertainty interval for a given value of the confidence probability P. Such an approach leads to error estimates from the upper bound (pessimistic estimation).

Besides this case, the measurement channel may include a device that provides the results of measuring physical values in a digital form. An example of such a device is a dynamometer, the output of which is the torque M_{sh}, defined as the tendency of a force to rotate a shaft about its axis, and revolution speed n_{sh} of its shaft. As a rule, measurements of those values in physical units are supplied by the dynamometer in digital form and go through the appropriate interface to the computer. Such devices have a metrological certificate from the manufacturer, in which the confidence interval Δ_X $(P = 0.95)$ of errors is determined. Therefore, the errors of measurement results may be simulated by pseudo-random numbers with a uniform distribution in the range $\mp \frac{\Delta_X}{2}$. Under the same uncertainty interval, the uniform distribution is characterized by the greatest variance, which makes it possible to obtain estimates of the errors from the upper bound.

Sequences $\tilde{y}_{i1}, \ldots, \tilde{y}_{in}$ $(i=1, m)$ obtained as a result of statistical modeling are subject to processing in order to get estimates of mathematical expectations of $\tilde{Y}_i$, their *rms* deviation $\hat{\sigma}_{\tilde{y}_i} = \hat{\sigma}_{\varepsilon_i}$ and the uncertainty interval $\Delta_{\tilde{Y}_i}$ $(P=0.95)$. To determine a value of the latter, an identification of the distribution law of E_i must be done. If the distribution law is not identified *a priori* (normal, uniform, etc.), then a histogram of ε_i is plotted and an empirical value of the uncertainty interval $\Delta_{\tilde{Y}_i}$ is determined.

The commonality of the Monte Carlo method makes it possible to implement the stated procedure of statistical tests as one of the operating modes of the *AMIS*. With the change of the operator A_c, only the object (software module for processing measurement results) of statistical tests will change.

The use of such a procedure can be considered in the example of turboprop engine tests. In this example, the purpose of the test was the experimental determination of its throttle response. The engine was tested according to the scheme with the attached dynamometer, which absorbed the engine power. This structure called a test bed (Fig. 3.6) is located in a test cell.

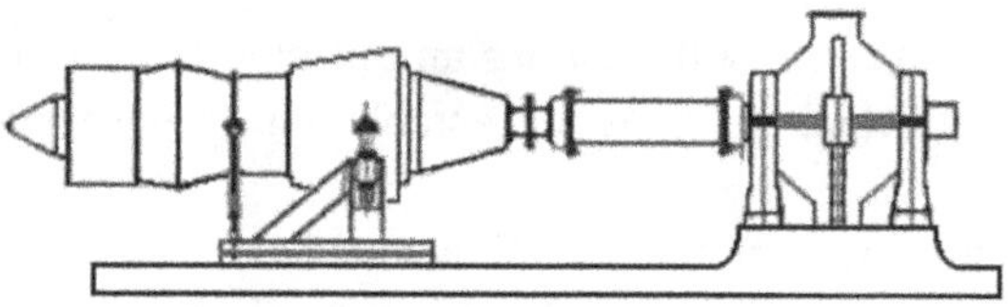

Fig. 3.6: Sketch of the test bed with the engine.

During the engine test a set of the following physical values was measured: V_f – volume of consumed fuel; ρ_f – fuel density; t_f- time of consumption of the volume V_f; P^*_{inlet} – the total pressure, T^*_{inlet} – stagnation (total) temperature, and ΔP_{inlet} – dynamic pressure of the inlet airflow; P_{noz} – static pressure at the edge of the exhaust nozzle; T^*_{EGT} – total exhaust gas temperature; M_{sh} – shaft torque; and n_{sh} – shaft revolution speed. Results of these measurements are used for the calculation of equivalent power N_e and corrected fuel consumption G_{fcor}. The experimental data for a sequence of given power points (designated as "O") is represented on Fig. 3.7.

In this case, the software module which was subjected to the statistical simulations, has the functional scheme represented in Fig. 3.8.

Besides N_e and G_{fcor}, this diagram represents the following calculated parameters: fuel consumption G_f, engine airflow G_{air}, shaft power N_{sh} and its corrected value N_{shcor} as well as the jet nozzle power N_{noz}.

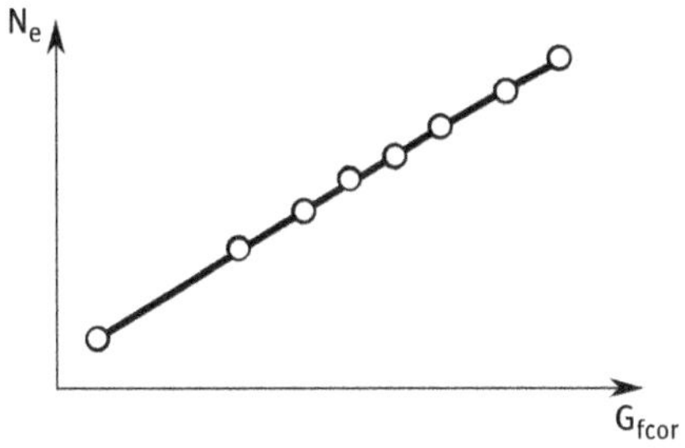

Fig. 3.7: Experimental throttle characteristic.

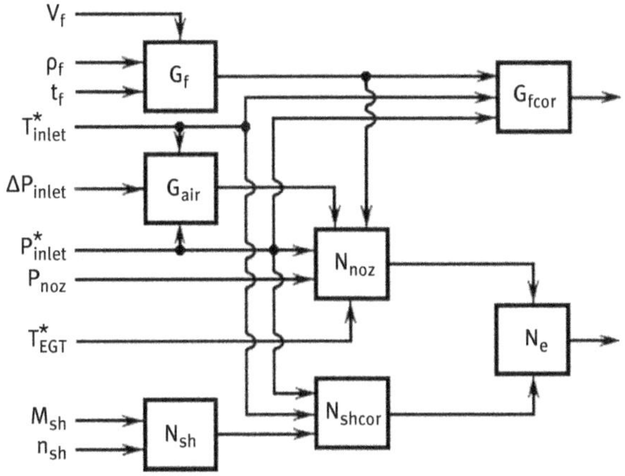

Fig. 3.8: Calculation scheme for N_e and G_{fcor}.

Metrological characteristics of measurement results were defined by the results of calibration procedures of a few channels as well as from metrological certificates of the remaining channels. Therefore, the simulation of errors of $\tilde{P}^*_{inlet}$, $\Delta\tilde{P}_{inlet}$, and $\tilde{P}_{noz}$ was done using pseudo-random numbers with a normal distribution. Other measurement errors were simulated by pseudo-random numbers with a uniform distribution. The results of statistical simulations with $n = 800$ are illustrated in Fig. 3.9.

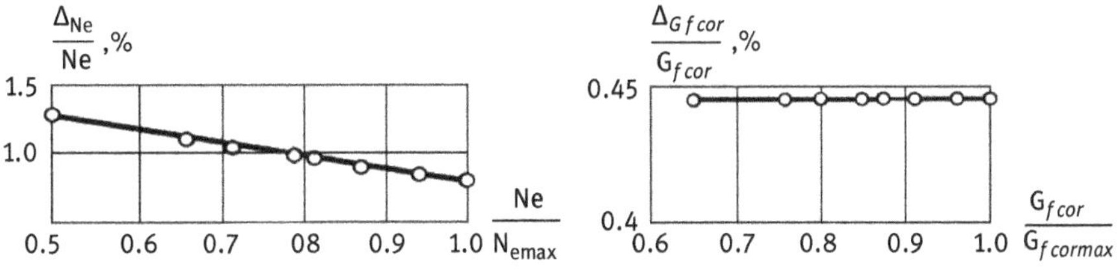

Fig. 3.9: Relative confidence intervals ($P = 0.95$) of test results.

Identification of the distribution of G_{fcor} estimates yields the normal law. In fact, the chi-square test gave the value of $\chi^2_{115} \approx 120$, and the critical value of this statistic at a significance level of $\alpha = 0.05$ is $\chi^2_{0.95,120} \approx 147$. As a result, the confidence interval $(P = 0.95)$ of $\tilde{G}_{fcor}$ was determined as $\Delta_{\tilde{G}_{fcor}} = 1.96.\hat{\sigma}_{\tilde{G}_{fcor}}$.

The distribution law of estimates of the equivalent power was different from the normal distribution and a histogram of scattered values $\tilde{N}_{ej}$ around their mathematical expectation $\hat{N}_e$ is represented in Fig. 3.10.

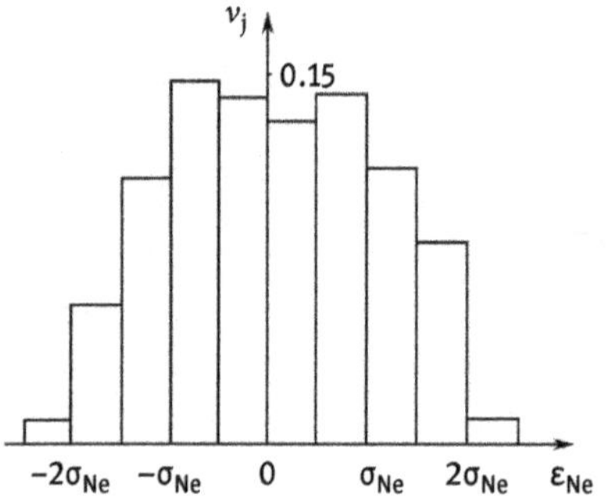

Fig. 3.10: Histogram of N_e estimates.

The empirical value of the confidence interval $(P = 0.95)$ of $\tilde{N}_e$ was $\Delta_{N_e} = 2\hat{\sigma}_{Ne}$.

As a rule, the characteristics of test objects are represented by monotonous curves without discontinuities. This fact allows to estimate errors of test results only for the minimum and maximum values of experimental characteristics.

3.2.2 Correlations of test results

The calculation diagram shown in Fig. 3.8 clearly demonstrates the fact that all measurement results used to calculate G_{fcor} are also involved in the calculation of N_e. This means that the matrix $\boldsymbol{A}_c$ is not diagonal, that is, some of its elements $a_{ij} \neq 0$. Recall that the covariance matrix of test results is defined as:

$$\boldsymbol{\Gamma}_{\tilde{Y}} = \boldsymbol{A}_c \boldsymbol{\Gamma}_{\tilde{X}} \boldsymbol{A}_c^{\mathrm{T}}.$$

Although, the covariance matrix $\boldsymbol{\Gamma}_{\tilde{X}}$ is diagonal, the matrix $\boldsymbol{\Gamma}_{\tilde{Y}}$ will not be diagonal. In other words, it will include covariances $\gamma_{\tilde{Y}_i \tilde{Y}_l} \neq 0$ indicating statistical dependencies of test results. The rate of statistical relationship between $\tilde{Y}_i$ and $\tilde{Y}_l$ is characterized by a value of the corresponding cross-correlation coefficient:

$$\rho_{\tilde{Y}_i \tilde{Y}_l} = \frac{\gamma_{\tilde{Y}_i \tilde{Y}_l}}{\sigma_{\tilde{Y}_i} \sigma_{\tilde{Y}_l}}.$$

Described in section 3.2.1, the approach for estimating the accuracy of test results was accomplished with estimates $\sigma_{\tilde{N}_e}$ and $\sigma_{\tilde{G}_{fcor}}$ being diagonal elements of the matrix $\mathbf{\Gamma}_{\tilde{Y}}$. In other words, only marginal variances of test results were taken into account. Such a situation is applicable in cases when accuracy of a given parameter Y_i is attributed only to its value without taking into account values corresponding to the remaining test results.

In fact, a point of an object's characteristics is determined by a set of parameters $Y_1, \ldots, Y_m$. This means that for each mode of the test object, the value $\tilde{Y}_i$ is accompanied by known estimates of $Y_1, \ldots, Y_{i-1}, Y_{i+1}, \ldots, Y_m$. Therefore, parameter $\tilde{Y}_i$ has to be characterized by a conditional *PDF*, and its errors – by conditional variance:

$$\sigma^2_{\tilde{y}_i/} = \sigma^2_{\tilde{y}_i/\tilde{y}_1,\ldots,\tilde{y}_{i-1},\tilde{y}_{i+1},\ldots,\tilde{y}_m} = \frac{|\mathbf{\Gamma}_{\tilde{Y}}|}{\left|\sigma^2_{\tilde{y}_i}\right|},$$

where $|\mathbf{\Gamma}_{\tilde{Y}}|$ is the determinant of matrix $\mathbf{\Gamma}_{\tilde{Y}}$, and $\left|\sigma^2_{\tilde{y}_i}\right|$ - the algebraic complement for the diagonal element $\sigma^2_{\tilde{y}_i}$ [5]. The complement is the determinant of a matrix derived from $\mathbf{\Gamma}_{\tilde{Y}}$ by crossing out i^{th} row and i^{th} column as well as multiplied by $(-1)^{i+i} = 1$.

The ratio of marginal and conditional variances may be demonstrated by a sample of two variables with a covariance matrix:

$$\mathbf{\Gamma}_{\tilde{Y}} = \begin{pmatrix} \sigma^2_{\tilde{y}_1} & \gamma_{\tilde{y}_1\tilde{y}_2} \\ \gamma_{\tilde{y}_2\tilde{y}_1} & \sigma^2_{\tilde{y}_2} \end{pmatrix}.$$

Taking into account the above expression for $\sigma^2_{\tilde{y}_i/}$, the conditional variance of the estimate $\tilde{Y}_1$ will be:

$$\sigma^2_{\tilde{y}_1/} = \frac{|\mathbf{\Gamma}_{\tilde{Y}}|}{\left|\sigma^2_{\tilde{y}_1}\right|} = \frac{\sigma^2_{\tilde{y}_1}\sigma^2_{\tilde{y}_2} - \gamma_{\tilde{y}_2\tilde{y}_1}\gamma_{\tilde{y}_1\tilde{y}_2}}{\sigma^2_{\tilde{y}_2}} =$$

$$\frac{\sigma^2_{\tilde{y}_1}\sigma^2_{\tilde{y}_2}\left(1-\rho^2_{\tilde{y}_1\tilde{y}_2}\right)}{\sigma^2_{\tilde{y}_2}} = \sigma^2_{\tilde{y}_1}\left(1-\rho^2_{\tilde{y}_1\tilde{y}_2}\right).$$

This implies that the value of the conditional variance is always smaller than the marginal value $\sigma^2_{\tilde{y}_1}$ when $\rho_{\tilde{y}_1\tilde{y}_2} \neq 0$. The effect of correlation can be visually illustrated by a sample of a characteristic represented in Fig. 3.11.

Test results corresponding to a given object mode are represented by the point $(\tilde{y}_1, \tilde{y}_2)$. These estimates have a joint density function $f(\tilde{y}_1, \tilde{y}_2)$ related, for instance, to the bivariate normal distribution. A *PDF* cross-section for $f(\tilde{y}_1, \tilde{y}_2) = 0.242$ provides the *rms* error ellipse laid inside a rectangle with the center $(\tilde{y}_1, \tilde{y}_2)$ and sides of $2\sigma_{\tilde{y}_1}$ and $2\sigma_{\tilde{y}_2}$. Recall that the *rms* ellipse limits a two-dimensional confidence $(P = 0.68)$ region for the point $(\tilde{y}_1, \tilde{y}_2)$.

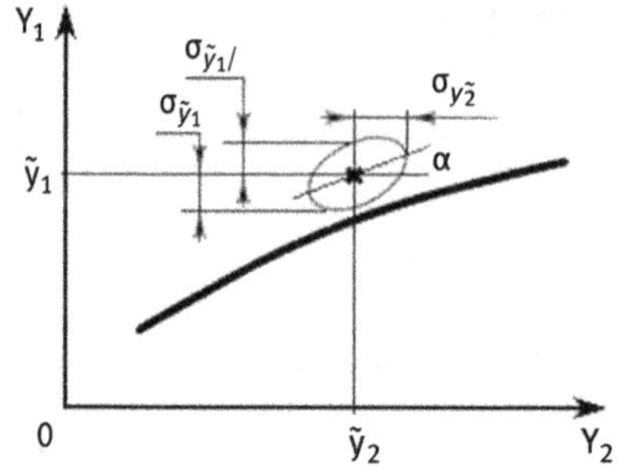

Fig. 3.11: Interpretation of a test result correlation.

The angle α of a major axis inclination is determined by the value of the cross-correlation coefficient:

$$tg2\alpha = \frac{2\rho_{\tilde{y}_1\tilde{y}_2}\sigma_{\tilde{y}_1}\sigma_{\tilde{y}_2}}{\sigma^2_{\tilde{y}_1} - \sigma^2_{\tilde{y}_2}}.$$

Varying $\rho_{\tilde{y}_1\tilde{y}_2}$ leads to variations of the angle a as well as to a deformation of the ellipse limited by borders of the above-mentioned rectangle. If there is a correlation between $\tilde{y}_1$ and $\tilde{y}_2$ $(\alpha \neq 0)$, the value of $\sigma_{\tilde{y}_1/}$ is always smaller than the value of $\sigma_{\tilde{y}_1}$. In other words, an increase in correlation increases this difference.

Therefore, in addition to diagonal elements of the matrix $\mathbf{\Gamma}_{\tilde{Y}}$, covariances $\gamma_{\tilde{y}_i\tilde{y}_l}$ $(i, l = \overline{1, m})$ need to be estimated using results of the simulation procedure. Its results contain the series $\tilde{y}_{i1}, \dots, \tilde{y}_{ij}, \dots, \tilde{y}_{in}$ and $\tilde{y}_{l1}, \dots, \tilde{y}_{lj}, \dots, \tilde{y}_{ln}$ that matches the bivariate probability density $f(\tilde{y}_i, \tilde{y}_l)$. Due to independence of each pair $(\tilde{y}_{ij}, \tilde{y}_{lj})$ from the previous and subsequent pairs, the logarithmic likelihood function may be written as:

$$l(\gamma_{\tilde{y}_i\tilde{y}_l}/\tilde{y}_{ij}, \tilde{y}_{lj}) = \ln L\left[\prod\nolimits_{j=1}^{n} f\left(\tilde{y}_{ij}, \tilde{y}_{lj}\right)\right] =$$

$$\sum\nolimits_{j=1}^{n} \ln\left[f\left(\tilde{y}_{ij}, \tilde{y}_{lj}\right)\right], j = \overline{1, n}.$$

In section 1.2.3 it was shown that the likelihood function asymptotically corresponds to the normal distribution law. By virtue of the fact that in the statistical simulation the value of $n = 800$, the expression for the function $l(\gamma_{\tilde{y}_i\tilde{y}_l}/\tilde{y}_{ij}, \tilde{y}_{lj})$ can be written in the following form:

$$l\left(\gamma_{\tilde{y}_j\tilde{y}_l}/\tilde{y}_{ij}, \tilde{y}_{lj}\right) \approx \sum_{j=1}^{n} \ln[(2\pi)^{-1}|\mathrm{H}|^{\frac{1}{2}} \times$$

$$\exp\left[-\frac{1}{2}\sum_{i=1}^{2} \cdot \sum_{l=1}^{2} \eta_{il}\left(\tilde{y}_{ij} - \mu_{\tilde{y}_i}\right)\left(\tilde{y}_{lj} - \mu_{\tilde{y}_l}\right)\right].$$

Here **H** is a matrix inverse to the covariance matrix $\mathbf{\Gamma}_{\tilde{y}_l\tilde{y}_i}$ which can be written as:

$$\mathbf{H}=\mathbf{\Gamma}_{\tilde{y}_l\tilde{y}_i}^{-1}=\begin{pmatrix}\sigma_{\tilde{y}_i}^2 & \gamma_{\tilde{y}_i\tilde{y}_l}\\ \gamma_{\tilde{y}_l\tilde{y}_i} & \sigma_{\tilde{y}_l}^2\end{pmatrix}^{-1}=\begin{pmatrix}\eta_{11} & \eta_{12}\\ \eta_{21} & \eta_{22}\end{pmatrix};$$

The determinant of this matrix is $|\mathrm{H}| = |\Gamma|^{-1}$.

A maximum of the $l(\gamma_{\tilde{y}_l\tilde{y}_i}/\tilde{y}_{ij},\tilde{y}_{lj})$ with respect to parameters η_{ij} corresponds to the following condition:

$$\frac{\partial l(\gamma_{\tilde{y}_l\tilde{y}_i}/\tilde{y}_{ij},\tilde{y}_{lj})}{\partial\eta_{il}}=0,$$

which will result in the following equation:

$$\sum_{j=1}^{n}\left[\frac{1}{|\mathrm{H}|}\cdot\frac{\partial|\mathrm{H}|}{\partial\eta_{il}}-\left(\tilde{y}_{ij}-\mu_{\tilde{y}_i}\right)\left(\tilde{y}_{lj}-\mu_{\tilde{y}_l}\right)\right]=0. \tag{4}$$

A derivative of the determinant |H| can be defined in the following form:

$$\frac{\partial|\mathrm{H}|}{\partial\eta_{il}}=\frac{\partial(\eta_{11}\eta_{22}-\eta_{21}\eta_{12})}{\partial\eta_{il}}=-\eta_{li},\ i\neq l.$$

Therefore, the expression (4) transforms to the following equality:

$$\sum_{j=1}^{n}\frac{\eta_{li}}{|\mathrm{H}|}\cdot(-1)^{i+l}=\sum_{j=1}^{n}\left(\tilde{y}_{ij}-\mu_{\tilde{y}_i}\right)\left(\tilde{y}_{lj}-\mu_{\tilde{y}_l}\right).$$

Bear in mind that the element η_{li} of the inverse matrix equals to the algebraic complement $\Gamma_{\tilde{y}_i\tilde{y}_l}$ to the element $\gamma_{\tilde{y}_i\tilde{y}_l}$ divided by the determinant $|\Gamma|$. In its tern, the complement $\Gamma_{\tilde{y}_i\tilde{y}_l}$ is the determinant of a matrix derived from $\mathbf{\Gamma}_{\tilde{y}_i\tilde{y}_l}$ by crossing out i^{th} row and l^{th} column as well as multiplied by $(-1)^{l+i}$.

As $\mathbf{H}^{-1}=\mathbf{\Gamma}$, the algebraic complement $\mathrm{H}_{il}=-\eta_{li}$ divided by the determinant |H| equals a corresponding element of matrix $\mathbf{\Gamma}$:

$$-\eta_{li}/|\mathrm{H}|=\gamma_{\tilde{y}_i\tilde{y}_l}.$$

Thus, the expression for calculating the covariance *MLE* of two random variables $\tilde{y}_i$ and $\tilde{y}_l$ can be determined as:

$$\hat{\gamma}_{\tilde{y}_i\tilde{y}_l}=\frac{\sum_{j=1}^{n}\left(\tilde{y}_{ij}-\mu_{\tilde{y}_i}\right)\left(\tilde{y}_{lj}-\mu_{\tilde{y}_l}\right)}{n}.$$

In fact, this expression is obtained from the condition of the maximum of the likelihood function with respect to the parameter η_{il}. Since the elements of matrices $\mathbf{\Gamma}_{\tilde{y}_l\tilde{y}_i}$

and **H** are functionally related, the invariance property of the *MLE*s allows to state that the obtained estimates $\hat{y}_{il}$ are the *MLE*s.

If instead of the unknown mathematical expectations $\mu_{\tilde{y}_i}$ and $\mu_{\tilde{y}_l}$ their estimates are used, then the $\hat{y}_{\tilde{y}_i\,\tilde{y}_l}$ expression is transformed to the following form:

$$\hat{y}_{\tilde{y}_i\tilde{y}_l} = \frac{\sum_{j=1}^{n}\left(\tilde{y}_{ij} - \hat{\mu}_i\right)\left(\tilde{y}_{lj} - \hat{\mu}_j\right)}{n-2}.$$

In turn, the *MLE* of the cross-correlation coefficient will be:

$$\hat{\rho}_{\tilde{y}_i\,\tilde{y}_l} = \frac{\hat{y}_{\tilde{y}_i\tilde{y}_l}}{\hat{\sigma}_{\tilde{y}_i}\hat{\sigma}_{\tilde{y}_l}}.$$

It should be noted that estimates of all parameters calculated using simulation results have some statistical uncertainty. Therefore, estimates of $\hat{\rho}_{\tilde{y}_l\,\tilde{x}_j}$ with small values should be verified for statistical significance. To this end, the null hypothesis is that the values of these cross-correlations are equal to zero $(H_1{:}\hat{\rho}_{\tilde{y}_l\,\tilde{x}_j} \neq 0)$. The competing hypothesis is written as $(H_1{:}\hat{\rho}_{\tilde{y}_l\,\tilde{x}_j} \neq 0)$.

Let us assume that the random variables for which the cross-correlation is estimated have a normal distribution. There is known the Fisher's *Z*-transformation of estimates $\hat{\rho}$ [6]:

$$T_n = \frac{\sqrt{n-3}}{2}\ln\left(\frac{1+\hat{\rho}}{1-\hat{\rho}}\right) = Z.$$

Therefore, the critical value of statistics T_n can be determined using the upper and lower bounds of the confidence interval for the standard normal variable *Z*:

$$z_{\alpha/2} \leq \frac{\sqrt{n-3}}{2}\ln\left(\frac{1+\hat{\rho}}{1-\hat{\rho}}\right) \leq z_{1-\alpha/2}.$$

As a result, the condition for rejecting null hypothesis can be written as:

$$|\hat{\rho}| > \rho_{1-\alpha} = \frac{\exp\left(2z_{1-\alpha/2}/\sqrt{n-3}\right) - 1}{\exp\left(2z_{1-\alpha/2}/\sqrt{n-3}\right) + 1}.$$

When this condition is met, the difference $\hat{\rho}$ from zero cannot be attributed to the statistical variability of estimates. In particular, for $n = 800$ and $\alpha = 0.05$, the value of $\rho_{0.95} = 0.07$.

An actual effect of test result correlation can be demonstrated by an example of experimental estimation of characteristics of a turbojet engine. One of its characteristics, called an altitude-speed, defines a value of an engine's thrust *R* (measured in newtons – *N*) as a function of a corrected fuel consumption G_{fcor} (measured in *kg/h*). Dependence of *R* on G_{fcor} is conditioned on values of flight speed *M* (measured in dimensionless Mach numbers) and altitude H (measured in meters). A typical view of altitude-speed characteristics for the "Maximal Rate" (*MR*) mode is presented in Fig. 3.12.

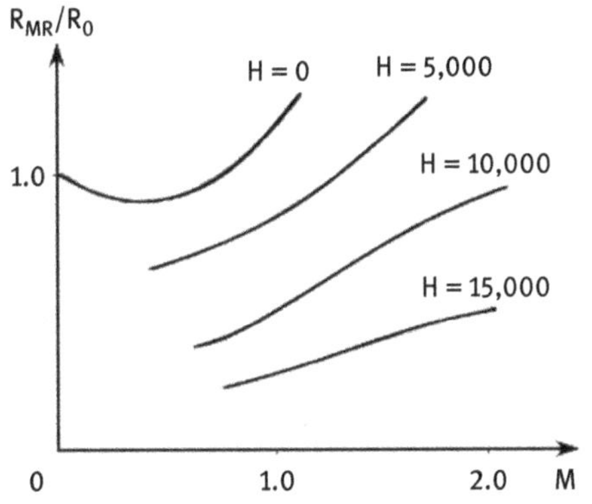

Fig. 3.12: Altitude-speed characteristics of the *GTE*.

Such characteristics are estimated by testing the *GTE* in an altitude chamber (Fig. 3.13).

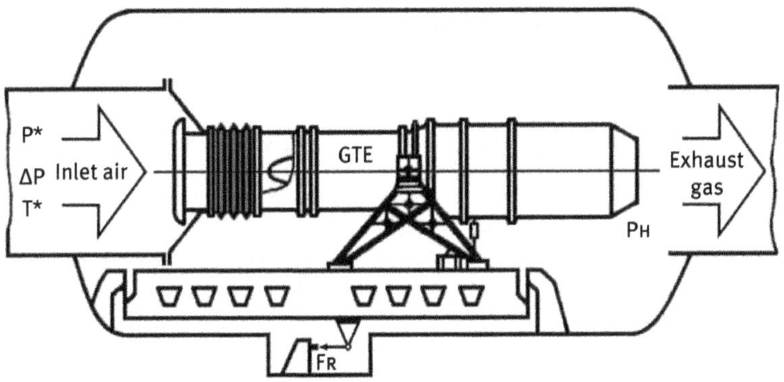

Fig. 3.13: *GTE* tests with a simulation of flight speeds and altitudes.

Flight altitude is determined by the value of static pressure P_H created by exhausting gas from the chamber. Flight speed is simulated by supplying the inlet air with values of the stagnation (total) pressure P^*, dynamic pressure ΔP, and total temperature T^* corresponded to the required values of M and H.

To estimate the value of thrust R, the *GTE* is mounted on a moving (dynamic) platform of the thrust measurement system (*TMS*). The dynamic platform is connected to the frame of the *TMS* through flexure plates, so that the thrust force F_R

applied to the dynamic platform can be measured. Results of measurements of F_R, inlet air pressures and P_H are used to calculate the value of engine thrust R.

The G_{fcor} are calculated using measured values of fuel mass flow G_f as well as P^* and T^*.

Therefore, any experimental point of the altitude-speed characteristics will be accompanied by the following covariance matrix:

$$\boldsymbol{\Gamma}_{\tilde{Y}} = \begin{pmatrix} \sigma^2_{\tilde{R}} & \gamma_{\tilde{R}\tilde{G}} & \gamma_{\tilde{R}\tilde{H}} & \gamma_{\tilde{R}\tilde{M}} \\ \gamma_{\tilde{G}\tilde{R}} & \sigma^2_{\tilde{G}} & \gamma_{\tilde{G}\tilde{H}} & \gamma_{\tilde{G}\tilde{M}} \\ \gamma_{\tilde{H}\tilde{R}} & \gamma_{\tilde{H}\tilde{G}} & \sigma^2_{\tilde{H}} & \gamma_{\tilde{H}\tilde{M}} \\ \gamma_{\tilde{M}\tilde{R}} & \gamma_{\tilde{M}\tilde{G}} & \gamma_{\tilde{M}\tilde{H}} & \sigma^2_{\tilde{M}} \end{pmatrix}.$$

The elements of this matrix can be estimated by using the results of simulating the measurement errors as was described in section 3.1.2. Applying this procedure to measurement results at the MR regime in given conditions $(M = \text{const}; H = \text{const})$ yielded the following matrix $\boldsymbol{\Gamma}_{\tilde{Y}}$:

$$\boldsymbol{\Gamma}_{\tilde{Y}} = \begin{pmatrix} 47 \cdot 10^2 & 74 & -250 & -0.32 \\ 74 & 4.8 & -0.84 & -93 \cdot 10^{-4} \\ -250 & -0.84 & 93 & 46 \cdot 10^{-3} \\ -0.32 & -93 \cdot 10^{-4} & 46 \cdot 10^{-3} & 48 \cdot 10^{-6} \end{pmatrix}.$$

The corresponding correlation matrix is

$$\mathbf{P}_{\tilde{Y}} = \begin{pmatrix} 1 & 0.49 & -0.39 & -0.68 \\ 0.49 & 1 & -0.04 & -0.61 \\ -0.39 & -0.04 & 1 & 0.69 \\ -0.68 & -0.61 & 0.69 & 1 \end{pmatrix}.$$

Theoretically, there is no correlation between only two parameters G_{fcor} and H because the correction values of G_f are calculated using only measurements of P^* and T^*. The non zero estimate $\hat{\rho}_{\tilde{G}\tilde{H}} = -0.04$ simply reflects a property of the simulation procedure.

Considerable ranks (−0.39 . . . 0.69) of rest correlations will noticeably lead to a difference of marginal and conditional variances of thrust estimates. In the examined case, the marginal *rms* error of R is equal to $\hat{\sigma}_{\tilde{R}} = 68.6N$, while the conditional error – only $\hat{\sigma}_{\tilde{R}/} = 49\,N$, that is, the marginal variance is bigger the conditional one in 1.4 times. The considered example shows that ignoring the correlations of test results makes the real accuracy of characteristic estimates less precise. This fact may lead to unjustified requirements to increase accuracy of measurement channels. In other words, this could translate to needless additional cost of tests.

3.3 Ensuring of required accuracy

Estimating the errors of test results is a typical direct analysis of a system when the characteristics of its outputs (errors of test results) are estimated from known inputs (measurement errors). Sometimes, results of such an analysis can reveal a discrepancy between the obtained errors of parameters estimates and requirements imposed on their accuracy. In this case, it will be necessary to solve the inverse task (the task of synthesis). Its solution must define requirements for the accuracy of measurement results based on the required accuracy of test results.

As a rule, synthesis seems to be a more difficult task due to its multifactor character. Consider such a task when it comes to ensuring accuracy of a parameter Y_l. In this case, there is a possibility to obtain its required accuracy by various combinations of errors in measurement results. In other words, this inverse task has a multivariate solution. Furthermore, it is quite possible to assume that one variant among the different combinations of measurement errors can be optimal in the sense of a certain criterion.

3.3.1 Ranking of error sources

To form requirements for measurement errors, a quantitative description of their link with the test result errors needs to be available. Such a description is provided by the metrological model of data processing (errors propagation) presented in the section 2.2:

$$\mathbf{E} = \boldsymbol{A}_c \Xi,$$

In accordance with this model, the link between the error ε_l and measurement errors ξ_i is described by the following expression:

$$\varepsilon_l = \sum_{i=1}^{k} a_{li} \cdot \xi_i, \tag{5}$$

where $a_{li} = \frac{\partial Y_l}{\partial X_i}|_{\tilde{X}}$ and $\tilde{X}$ are estimates of measured values at a given mode of the test object.

Let us turn to the cross-covariance $\gamma_{\varepsilon_l \xi_j}$ of variables ε_l and ξ_j:

$$\gamma_{\varepsilon_l \xi_j} = M\left[\varepsilon_l \cdot \xi_j\right] = M\left[\left(\sum_{i=1}^{k} a_{li}\xi_i\right)\xi_j\right] =$$

$$M\left[a_{lj}\xi_j^2\right] + \sum_{\substack{i=1 \\ i \neq j}}^{k} a_{li} M[\xi_i \cdot \xi_j] = a_{lj}\, \sigma_{\xi_j}^2,$$

The remaining cross-covariances $M[\xi_i\xi_j] = 0$ due to independence of measurement results. Therefore, the cross-correlation $\rho_{\varepsilon_l\xi_j}$ will be equal:

$$\rho_{\varepsilon_l\xi_j} = \frac{\gamma_{\varepsilon_l\xi_j}}{\sigma_{\varepsilon_l}\sigma_{\xi_j}} = \frac{a_{lj}\,\sigma^2_{\xi_j}}{\sigma_{\varepsilon_l}\sigma_{\xi_j}} = \frac{a_{lj}\,\sigma_{\xi_j}}{\sigma_{\varepsilon_l}}.$$

In other words, the value of $\rho_{\varepsilon_l\xi_m}$ determines the relative part of the *rms* error of Y_l which is due to errors of X_j. This fact justifies the use of $\rho_{\varepsilon_l\xi_j}$ as a quantitative measure of the links between the errors of the measured and calculated values.

Another reason for the use of $\rho_{\varepsilon_l\xi_j}$ is associated with the simplicity of its evaluation. Indeed, the results of statistical simulation of measurement errors include series of both calculated ($\tilde{\boldsymbol{Y}}$) and measured ($\tilde{\boldsymbol{X}}$) values. Therefore, the cross-correlation coefficients can be estimated as:

$$\hat{\rho}_{\tilde{y}_l\tilde{x}_j} = \frac{\hat{\gamma}_{\tilde{y}_l\tilde{x}_j}}{\hat{\sigma}_{\tilde{y}_l}\hat{\sigma}_{\tilde{x}_j}} = \frac{\sum_{i=1}^{n}\left(\tilde{y}_{li} - \hat{\mu}_{\tilde{y}_l}\right)\left(\tilde{x}_{ji} - \hat{\mu}_{\tilde{x}_j}\right)}{\sqrt{\sum_{i=1}^{n}(\tilde{y}_{li} - \hat{\mu}_{\tilde{y}_l})^2 \cdot \sum_{j=1}^{n}(\tilde{x}_{ji} - \hat{\mu}_{\tilde{x}_j})^2}}.$$

The approach for ranking sources of measurement errors can be illustrated by the example of tests of an axial compressor stage. A scheme of a corresponding test rig is represented in Fig. 3.14.

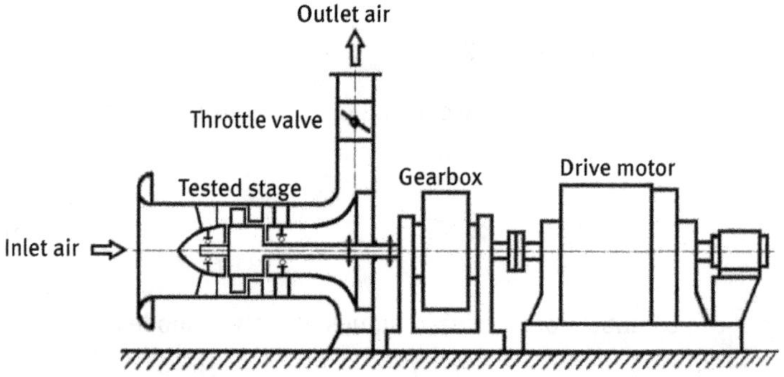

Fig. 3.14: Scheme of a test rig for a single axial stage.

The single axial stage has a rotating disk (work wheel) with airfoils (blades) set on its rim. The air stream is pumped (accelerated) by the blades in the rearward direction since the work wheel is driven by an electrical motor via a gearbox. The gearbox is used to multiply the rotation speed of the work wheel. The stage rotor is mounted with a set of front and rear bearings that support its position.

There are two rows of stationary airfoils (guide vanes) fixed to the stage stator. These inlet and outlet guide vanes, convert the kinetic energy of air into pressure energy. In addition, they parry a tendency of the air flow to move together with the rotating wheel, that is, re-direct it axially.

The characteristics of the stage is determined by the rotation speed n of its shaft and the stage mass air flow G_{air} controlled by the throttle valve. The purpose of the tests is to estimate the stage efficiency characteristics. Its experimental values are presented in Fig. 3.15.

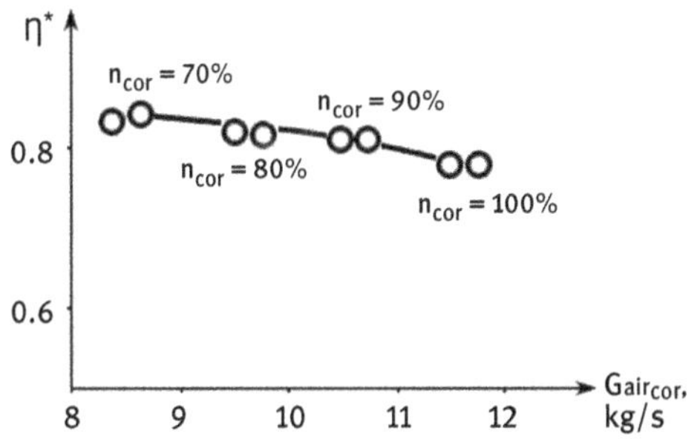

Fig. 3.15: Line of stage efficiency maximums.

This chart demonstrates a plot of maximal values of the stage efficiency η^* against corrected values of G_{air} for given values of n_{cor}. The stage efficiency characterizes the transfer of mechanical energy of the blades to the compressed air. There are two points with highest values of η^* for a given value of shaft speed.

During the tests, the following values in addition to n were measured:

- balancing torque M_b of the gearbox stator by means of a moment arm acting on a load cell;
- pressures ΔP, P^*, and temperatures $T^*$of air streams along the radii of the inlet and outlet cross-sections of the stage;
- lubricant oil flow G_{oil} as well as its inlet and outlet temperatures T_{oil}^* for the stage bearings and the gearbox.

The statistical simulation of the errors of the measurements listed above was done at one point of n_{cor} by pseudorandom numbers with a uniform distribution. The relative confidence intervals ($P = 0.95$) of measurement results had the following values: $\Delta_{Mb}/M_b = 0.5\%$ (FS); for all kinds of pressures $\Delta_P/P = 0.3\%$ (FS); $\Delta_n/n = 0.01\%$ (FS); $\Delta_{Goil}/G_{oil} = 2.0\%$ (FS), and $\Delta_{T^*} = 1°C$ (for air and oil). The results of statistical simulation are presented in Fig. 3.16.

The presented confidence intervals ($P = 0.95$) correspond to marginal distributions of estimates of η^* and $G_{air_{cor}}$.

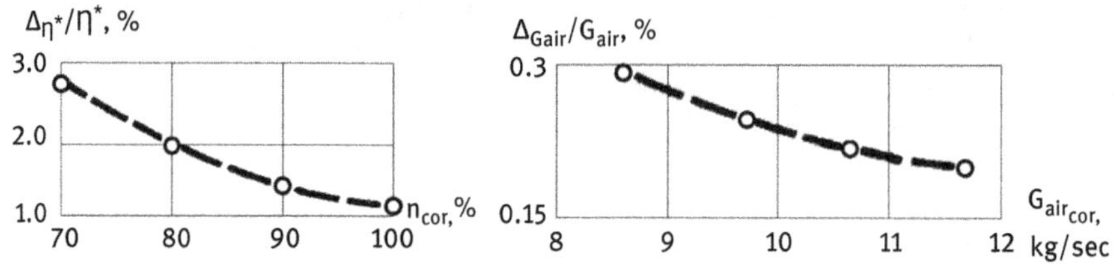

Fig. 3.16: Interval errors of η^* and $G_{air_{cor}}$ estimates.

It should be noted that more than two dozen measurement channels used for testing the compressor stage consist of six types of channels for measurements of:
- balancing torque of the gearbox,
- pressure of the air flow,
- temperature of the air flow,
- temperature of oil,
- flow rate of oil,
- and shaft rotation *n*.

This fact makes it possible to abandon the consideration of the errors of the measuring channels and to reduce the problem to the consideration of errors of values M_b, P, T^*, T_{oil}^*, G_{oil}, and n.

Taking into account (5), the following expression for the variance of the error E_l can be written down as follows:

$$\sigma_{\varepsilon_l}^2 = \sum_{i=1}^{k} a_{li}{}^2 \sigma_{\xi_i}^2.$$

It can be modified by dividing the left and right parts of this equality by $\sigma_{\varepsilon_l}^2$.:

$$1 = \sum_{i=1}^{k} \frac{a_{li}{}^2 \sigma_{\xi_i}^2}{\sigma_{\varepsilon_l}^2} = \sum_{i=1}^{k} \rho_{\varepsilon_l \xi_i}^2. \tag{6}$$

Therefore, correlation coefficients $\rho_{\varepsilon_l \xi_i}$ of k_i identical channels may be combined in one $\rho_{\tilde{Y}_l \tilde{X}_i}$:

$$\rho_{\tilde{Y}_l \tilde{X}_i}^2 = \sum_{j=1}^{k_i} \rho_{\varepsilon_l \xi_{ij}}^2.$$

This approach allows to reduce the number of $\rho_{\tilde{Y}_l \tilde{X}_i}$ taken into account.

For the example under consideration, expression (6) is converted to the following form:

$$\sum_{i=1}^{6} \rho_{\tilde{\eta}^* \tilde{X}_i}^2 = 1.$$

Bear in mind that the sum $\sum_{i=1}^{6} k_i = k$ is a total number of measurement channels whose readouts are used to calculate $\tilde{\eta}^*$.

Analysis of the statistical simulation results demonstrates that the dominant sources of the $\tilde{\eta}^*$ errors are measurements of the balancing torque and oil temperature (Fig. 3.17).

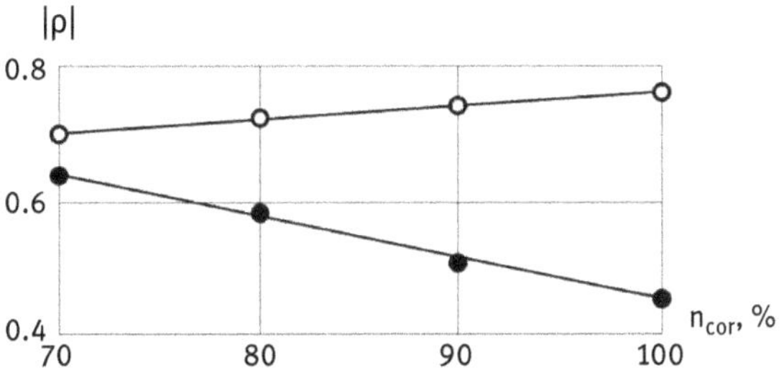

Fig. 3.17: Dominating cross-correlation coefficients. ("○" – $\hat{\rho}_{\tilde{\eta}^* \tilde{M}_b}$, "●" – $\hat{\rho}_{\tilde{\eta}^* \tilde{T}^*_{oil}}$)

For the mode $n_{cor} =$ 70%, values of the cross-correlation coefficients are $\hat{\rho}_{\tilde{\eta}^* \tilde{M}_b} = 0.71$, and $\hat{\rho}_{\tilde{\eta}^* \tilde{T}^*_{oil}} = 0.63$. The considerable effect of the errors in measuring the temperatures T^*_{oil} is explained as follows. Measurements of inlet and outlet oil temperature of the gearbox as well as stage bearings are used to estimate the energy losses due to mechanical friction. At $n =$ 70%, the relative proportion of losses is large, and an error of 1% leads to an error of estimating the value of torque on the stage shaft of more than 1%.

The fact that $\hat{\rho}^2_{\tilde{\eta}^* \tilde{M}_b} + \hat{\rho}^2_{\tilde{\eta}^* \tilde{T}^*_{oil}} \approx 0.5 + 0.4 = 0.9$ allows to exclude from consideration all other measurements due to their insignificant influence on the accuracy of estimates $\tilde{\eta}^*$.

3.3.2 Optimization of accuracy requirements

The task of ensuring the required accuracy of estimates of Y_l for a given mode of the test object can be formulated as follows. We take the relative error δ_l for the accuracy indicator of $\tilde{Y}_l$ defined as:

$$\delta_l = \Delta_{\tilde{Y}_l} / \hat{\mu}_{\tilde{Y}_l}.$$

Here $\Delta_{\tilde{Y}_l}$ is the confidence interval $(P = 0.95)$ of the estimates of Y_l, $\hat{\mu}_{\tilde{Y}_l} = \sum_{j=1}^{n} \tilde{y}_{lj} / n$. It is required to determine the accuracies of measurement results that will reduce the value of δ_l by Θ times.

Earlier, a model of error propagation for the value of Y_l was represented in the following form:

$$\sigma^2_{\tilde{Y}_l} = \sum_{i=1}^{k} a^2_{li}\sigma^2_{\tilde{X}_i}.$$

We assume that the distribution law of the errors does not depend on the value of the variances. Therefore, decreasing the accuracy of δ_l by Θ times means decreasing the variance $\sigma^2_{\tilde{Y}_l}$ by Θ^2 times. As a result, to ensure the accuracy of estimates of the parameter Y_l the following equation can be used:

$$\left(\frac{\sigma_{\tilde{Y}_l}}{\Theta}\right)^2 = \sum_{i=1}^{k}\left(\frac{a_{li}\sigma_{\tilde{X}_i}}{\theta_i}\right)^2,$$

where Θ_i is corresponding requirement to decrease the measurement error of $\tilde{X}_i$. Dividing both sides of this equality by $\sigma^2_{\tilde{Y}_l}$ and taking into account that $\rho_{\tilde{Y}_l\tilde{X}_i} = a_{li}\sigma_{\tilde{X}_i}/\sigma_{\tilde{Y}_l}$, one will get:

$$\sum_{i=1}^{k}\frac{\rho^2_{\tilde{y}_l\tilde{X}_i}}{\theta_i^2} = \frac{1}{\Theta^2}. \tag{7}$$

The next step is to define the criterion that allows to determine the optimal values of θ_i, that is, an optimal solution of the formulated task. As a rule, the practical activity of mankind occurs with limited resources. Therefore, the costs of achieving the goal are looked at as a universal criterion that may be applicable to the task of ensuring the required accuracy of the test results. Indeed, an increase in the accuracy is normally achieved be using more expensive but more accurate sensors, increasing the number of measurement channels, and so on. All these actions are associated with corresponding costs.

In each particular experiment, certain values of accuracy indicators are already provided, hence the cost of their subsequent improvement may be different. In addition, the errors of measurement results make various contributions to the magnitude of the errors of test results. Taking into account these facts, one can assume the existence of optimal (in terms of cost) requirements for measurement accuracies.

The complexity of mentioned approach is due to requirements for an available analytical description of costs as a function of θ_i. If a formal description of this inverse task is done, one can use the methods of the calculus of variations [7] to find an optimal solution of the task.

It should be noted that it is almost impossible to solve the task of ensuring the highest accuracy of test results for a given budget. In particular, this is due to the methodological difficulties of accurate calculating the cost to improve the accuracy of measurement results. Therefore, we will consider a different formulation of the inverse task – ensuring the required accuracy of the test results with minimal cost.

At the same time, the quantitative link of cost with an increase in the accuracy is replaced by a heuristic postulate: the higher the accuracy the greater the cost necessary to achieve it. In this way, the optimal condition of the considered inverse task may be written as:

$$\sum_{i=1}^{k} w_i\theta_i \underset{\theta_i}{\Rightarrow} min.$$

Here w_i represents the weight factors defined by experts for the comparability of cost required to improving the accuracy of heterogeneous measurements. In addition, the values of θ_i are related by a condition defined by expression (7). Thus, the task of ensuring the required accuracy of estimates of parameter Y_l is formulated as a conditional (constrained) optimization task. Its solution can be found by the method of Lagrange multipliers. A new variable $\lambda > 0$ (Lagrange multiplier) is introduced in order to compose the Lagrangian function $L(\theta_i, \lambda)$:

$$L(\theta_i, \lambda) = \sum_{i=1}^{k} \left(w_i\theta_i + \lambda \frac{\rho^2_{\tilde{Y}_l\tilde{X}_i}}{\theta_i^2} \right) \underset{\theta_i}{\Rightarrow} min.$$

A stationary point (local extremum) of this function corresponds to the first partial derivatives that are equal zero. As the second partial derivatives of $L(\theta_i, \lambda)$

$$\frac{\partial^2 L(\theta_i, \lambda)}{\partial \theta_i^2} = \frac{6\lambda\rho^2_{\tilde{Y}_l\tilde{X}_i}}{\theta_i^4} > 0,$$

the stationary point is a minimum.

The differentiation of the Lagrangian function yields a system of k equations:

$$\frac{\partial L(\theta_i, \lambda)}{\partial \theta_i} = w_i - 2\lambda \frac{\rho^2_{\tilde{Y}_l\tilde{X}_i}}{\theta_i^3} = 0.$$

Solving this system together with the constraint eq. (7) yields the following expression of θ_i:

$$\theta_i = \frac{2\Theta}{w_i} \cdot \sqrt[3]{\frac{w_i^2\rho^2_{\tilde{Y}_l\tilde{X}_i}}{4}} \cdot \sqrt{\sum_{j=1}^{k} \sqrt[3]{\frac{w_j^2\rho^2_{\tilde{Y}_l\tilde{X}_j}}{4}}}.$$

This expression allows to calculate a required increase in measurement accuracies that ensures a decrease of error of Y_l by Θ times in an optimal way.

This approach can be illustrated by the example discussed in section 3.1.1 – tests of an axial compressor stage. When analyzing test results, a conditional confidence interval of the estimate $\tilde{\eta}^*$ was used as an accuracy indicator. Due to the known relationship between the variances of the marginal and conditional distributions, a requirement to increasing the accuracy of $\tilde{\eta}^*$ was reduced to the marginal distributions in a form of the following equality:

$$\frac{\Delta_{\eta^*}}{\tilde{\eta}^*} = 1.5\%, \qquad n_{cor} = 70\%.$$

In others words, the accuracy of estimates of stage efficiency has to be increased by $\Theta_{n=70\%} = 2.7/1.5 = 1.8$ times.

It was already determined that only measurement results of M_b and T^*_{oil} need to be taken into account because the remaining measurements have insignificant influence on the accuracy of estimates $\tilde{\eta}^*$. Actually, the sum of corresponding $\hat{\rho}^2_{\tilde{\eta}^*\tilde{X}_i}$ evaluated from the condition (6) is:

$$\sum_{i=1}^{4} \hat{\rho}^2_{\tilde{\eta}^*\tilde{X}_i} = 1 - \hat{\rho}^2_{\tilde{\eta}^*\tilde{M}_b} - \hat{\rho}^2_{\tilde{\eta}^*\tilde{T}^*_{oil}} = 1 - 0.9 = 0.1.$$

As a result, the inverse task is reduced to a two-factorial problem:

$$\left(\frac{\hat{\rho}_{\tilde{\eta}^*\tilde{M}_b}}{\theta_{M_b}}\right)^2 + \left(\frac{\hat{\rho}_{\tilde{\eta}^*\tilde{T}^*_{oil}}}{\theta_{T^*_{oil}}}\right)^2 = \frac{1}{1.8^2} - 0.1 = \frac{1}{2.2^2} = \frac{1}{\Theta^2_m},$$

where $\Theta_m = 2.2$ is a modified value of the parameter $\Theta_{n=70\%}$.

The weight coefficients were defined as $w_{M_b} = 10$ and $w_{T^*_{oil}} = 1$. Increasing the accuracy of measurements of the balancing torque appears more costly. In the described test, the temperature of the oil is measured by a single thermocouple in the inlet and outlet positions. Using their individual calibrations or replacing them with resistor thermometers (*RTDs*) will provide improved accuracy in a simple way.

The above expression of θ_i allows to obtain the following values of $\theta_{\tilde{M}_b}$ and $\theta_{\tilde{T}^*_{oil}}$:

$$\theta_{M_b} = \frac{2\cdot\Theta_m}{w_{M_b}} \cdot \sqrt[3]{\frac{w^2_{M_b}\cdot\hat{\rho}^2_{\tilde{\eta}^*\tilde{M}_b}}{4}} \cdot \sqrt{\sqrt[3]{\frac{w^2_{M_b}\cdot\hat{\rho}^2_{\tilde{\eta}^*\tilde{M}_b}}{4}} + \sqrt[3]{\frac{w^2_{T^*_{oil}}\cdot\hat{\rho}^2_{\tilde{\eta}^*\tilde{T}^*_{oil}}}{4}}} =$$

$$\frac{2\cdot 2.2}{10}\cdot\sqrt[3]{\frac{10^2\cdot 0.5}{4}}\cdot\sqrt{\sqrt[3]{\frac{50}{4}}+\sqrt[3]{\frac{0.4}{4}}} = 1.7;$$

$$\theta_{T^*_{oil}} = \frac{2\cdot 2.2}{1}\cdot\sqrt[3]{\frac{0.4}{4}}\cdot\sqrt{\sqrt[3]{\frac{50}{4}}+\sqrt[3]{\frac{0.4}{4}}} = 3.4$$

Thus, the measurements of M_b with relative error of 0.3% (*FS*) and T_{oil}^* with error of 0.3°*C* will provide estimates of efficiency η^* with relative error of 1.5%.

For the stage mode $n = 100\%$, values of cross-correlation coefficients are $\hat{\rho}_{\tilde{\eta}^* \tilde{M}_b} = 0.77$ and $\hat{\rho}_{\tilde{\eta}^* \tilde{T}^*_{oil}} = 0.46$. So, a sum of the remainder $\hat{\rho}^2_{\tilde{\eta}^* \tilde{X}_i}$ is:

$$\sum_{i=1}^{4} \hat{\rho}^2_{\tilde{\eta}^* \tilde{X}_i} = 1 - 0.77^2 - 0.46^2 = 0.19.$$

The above indicated accuracy of M_b and T_{oil}^* measurements will provide an increase in the accuracy of estimates $\tilde{\eta}^*$ by $\Theta_{n=100\%}$ times:

$$\Theta_{n=100\%} = \frac{1}{\sqrt{\left(\frac{0.77}{1.7}\right)^2 + \left(\frac{0.46}{3.4}\right)^2 + 0.19}} \approx 1.5.$$

This implies that the values of efficiency will be estimated with errors of $1.5\% \le \frac{\Delta_{\tilde{\eta}^*}}{\tilde{\eta}^*} \le 0.8\%$ for modes of the stage in the range of $70\% \le n_{cor} \le 100\%$.

3.3.3 Multiple measurements

Since the statistical uncertainty of the average value of n measurements decreases by a factor of $\sqrt{n}$, multiple measurements are an effective means of increasing the accuracy of test results. Moreover, the *AMIS* provides practically unlimited abilities in terms of acquisition, storing, and averaging measurement results for the implementation of this approach. Usually, $n = 10...25$, which ensures a reduction of the statistical uncertainty of the measurement results by 3 to 5 times.

At the same time, using multiple measurements is based on another reason as well. The work of modern engineering products is controlled by the automatic control systems (*ACSs*). They provide the stability of the object's parameters in steady-state modes and required dynamic characteristics of objects during their operation in transient regimes.

In steady-state modes, the *ACS* offsets the action of both external disturbances caused by the operating conditions of the test object and internal ones associated with physical phenomena inherent in the working process of the object. The *ACS* maintains the stability of one or several controlled parameters by changing a few controlling factors (physical values) that determine the nature of the object workflow. Like any technical device, the *ACSs* perform their job with some kind of efficiency. In other words, the stability of the parameters of the object is maintained with some level of accuracy. Hence, the magnitudes of physical values measured during tests will fluctuate within the limits determined by this accuracy. A frequency of this fluctuation is determined by the so-called object time constant τ which characterizes the inertial lag in the change of the controlled parameters after increasing or decreasing the values of controlling factors.

Referring to the *GTE*, where the main parameter to be controlled is the revolution speed of the rotor shaft. Its controlling factor is the fuel supply to the combustion chamber. The engine time constant τ characterizes the inertial lag in the change of the shaft speed when the fuel supply changes. This lag is mainly due to the moment of inertia of the rotor, in other words, depends on its mass and diametric dimensions. The larger they are, the longer the energy must be supplied after increasing the fuel supply in order for the rotor speed to change. Usually, fluctuations of measured values of the *GTE* look similar to wave oscillations with infra-low frequencies in the range of 0.5 … 1.5 *Hz*.

It should be noted that modern measurement channels utilize the "sample & hold" technique which provides measurement results related to the same point in time. The period of channel sampling is irrelevant when it comes to reduction of statistical uncertainties related instrumentation errors. Consequently, the next sampling cycle can be started immediately after acquisition of preceding data.

As far as the fluctuations of measured values are concerned, the interval Δt between measurement cycles determines the time basis of averaging the oscillations. For example, Fig. 3.18 shows the results of multiple measurements of infra-low fluctuations for two values of the time interval – Δt_1 and $\Delta t_2 (\Delta t_2 = 10\Delta t_1)$.

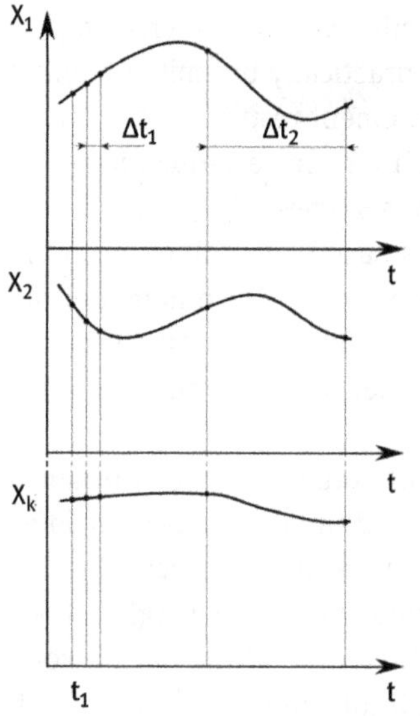

Fig. 3.18: Time diagram of multiple measurements.

The diagram clearly demonstrates that a small value of Δt does not provide sufficient representation in terms of the observed magnitude of oscillating values X_i. Therefore, the choice of the interval between the measurement cycles must be correlated with the time constants of test objects. Moreover, Δt may vary somewhat from one measurement cycle to another. Such an approach will randomize sampling the fluctuating values that are considered quasi-deterministic on a limited time base.

It should be noted that the fluctuation of a controlled parameter may greatly affect the magnitudes of the remaining measured values. For instance, reducing the rotation speed of the *GTE* by 1% may lead to a decrease of engine thrust by at least 3%. The engine *ACS* called the Engine Control Unit (*ECU*) supports rotor speeds with errors of ≤ 0.2%. It turns out that the work of the *ECU* may cause fluctuations of thrust values in the range of 0.6%. At the same time, the requirements for instrumental errors of the thrust measurement system stipulate their values to be ≤ 0.5%. Therefore, when a particular object is tested, the ratio of instrumental errors (σ_{ch}) and fluctuations (σ_{fl}) of the measured values should be evaluated. This allows to verify the significance of fluctuations, that is, to verify their proportionality with respect to measurement uncertainties.

In the case of fluctuations of measured values X_i, their variances can be represented in the following form:

$$\sigma_{X_i}^2 = \sigma_{ch_i}^2 + \sigma_{fl_i}^2.$$

The variance $\sigma_{X_i}^2$ should be estimated in modes "*IR*" and "*MR*". If $\hat{\sigma}_{\bar{X}_i}/\hat{\sigma}_{ch_i} > 1.15$ (equivalent of $3\hat{\sigma}_{fl_i} < \hat{\sigma}_{ch_i}$), fluctuations of measured values X_i cannot be ignored. It means that instead of $\hat{\sigma}_{ch_i}$, the value of the estimate $\hat{\sigma}_{X_i}^2$ may be used in the Monte Carlo procedure simulating uncertainties of measurement results. However, this issue is methodical and cannot be applied to the subject matter of this book, which considers only instrumental errors of measurement channels. Therefore, multiple measurements are considered in this book as effective means of increasing the accuracy of test results.

4 Transient regimes in test objects

In addition to the steady-state modes of operation, the test objects are also tested at transient regimes: startup, transitions from one steady-state mode to another, and shutdown. Such tests allow to estimate the dynamic properties of the test object which determine its behavior by observing changes of the modes and conditions of operation. According to this task, the measurement channels of the *AMIS* provide not only estimates of physical values but also their changes over time. In other words, the channels must measure dynamic processes $X_i(t)$ that occur in the test object. Therefore, in addition to the accuracy indicators discussed in the previous chapter, the characteristics of inertial properties of the *MC*s should also be considered. These studies can be performed using approaches and methods of the theory of dynamical systems.

4.1 Processes in dynamic systems

A dynamic system is any object or process whose state is defined by a set of variables at some point in time and for which there is a rule that describes the evolution of the initial state over time. If for a given moment of time only one future state follows from the current state, then the system is deterministic.

4.1.1 Deterministic processes

A system is called linear if, under the influence of an input process (disturbance) $x_{in}(t)$, its state $x_{out}(t)$ at the moment of time t is determined by the following convolution equation:

$$\int_0^t h(t-\tau)x_{in}(\tau)d\tau = \int_0^t h(\tau)x_{in}(t-\tau)d\tau = x_{out}(t).$$

Here $h(t)$ is a certain function, assumed to be piecewise continuous and it is called the weight function of the system. This function $h(t)$ describes the state of the system at the moment t after exposure to Dirac's delta (impulse) function $\delta(t-t_0)$ from a state of rest. The delta function has zero width and infinite height:

$$\delta(t-t_0) = \begin{cases} +\infty, & t = t_0 \\ 0, & t \neq t_0 \end{cases};$$

https://doi.org/10.1515/9783110666670-005

its integral (area) equals 1:

$$\int_{-\infty}^{\infty} \delta(t - t_0)dt = 1.$$

Another name for $h(t)$ is the impulse-response function (*IRF*).

The weight function defines the influence of the system input at the moment $(t - \tau)$ on its output at time t. Such a linear system is called time-invariant because the weight function is a function of the time difference. A physically realizable system should only respond to previous values of the input $x_1(t)$, that is, the following condition must be met:

$$h(t) = 0, \quad t < 0.$$

A system is said to be stable, if with any input limited in value, its output is also limited in its value. In other words, a stable linear system must have an absolutely integrable weight function:

$$\int_{-\infty}^{\infty} h(t)dt < \infty.$$

In the frequency domain, a linear dynamic system can be characterized by the frequency response function (*FRF*) $H(\omega)$. The *FRF* is defined as the Fourier transform of the weight function:

$$H(\omega) = \int_{-\infty}^{\infty} h(t) \cdot e^{-i\omega t}dt,$$

where $\omega = 2\pi f$ is the angular frequency, f – the cyclic frequency, and $i = \sqrt{-1}$. So, the *FRF* is a complex function that has two equivalent representations:

$$H(\omega) = U(\omega) + i \cdot V(\omega) = A(\omega) \cdot e^{i\Phi(\omega)}.$$

$U(\omega)$ and $V(\omega)$ are its real and imaginary parts, respectively; $A(\omega)$ and $\Phi(\omega)$ are the magnitude (amplitude) and angle (phase) components of the vector representation of the *FRF*. The relationships between these two forms of the $H(\omega)$ is described by the following formulas:

$$A(\omega) = \sqrt{U(\omega)^2 + V(\omega)^2};$$

$$\Phi(\omega) = tan^{-1}\left[\frac{V(\omega)}{U(\omega)}\right].$$

The frequency response function can be given the following physical interpretation. Let the input of a linear system be a harmonic oscillation of the following form:

$$X_{in}(t) = A_1 \sin(\omega t + \varphi),$$

and its second derivative is:

$$\frac{d^2 x_{in}(t)}{dt^2} = -\omega^2 \cdot x_{in}(t).$$

Taking into account that:

$$\frac{d^n x_{out}(t)}{dt^n} = \int_0^t h(\tau) \frac{d^n x_{in}(t-\tau)}{dt^n} d\tau,$$

one can write:

$$\frac{d^2 x_{out}(t)}{dt^2} = \int_0^t h(\tau) \cdot (-\omega^2) \cdot x_{in}(t-\tau)\, d\tau = -\omega^2 \cdot x_{out}(t).$$

This means that the output process of the system is also harmonic. Consequently, the ratio of the amplitudes of the input and output processes determines the $A(\omega)$ component, and the difference of their phases determines the $\Phi(\omega)$ component.

Note that the convolution of two functions in the time domain corresponds to the multiplication of the Fourier images of these functions. As a result, the frequency spectrum of the output process of a linear system can be defined as the product of the input spectrum and the system's *FRF*.

An example of a linear system is a system described by an ordinary differential equation with constant coefficients:

$$a_0 \frac{d^n x_{out}(t)}{dt^n} + a_1 \frac{d^{n-1} x_{out}(t)}{dt^{n-1}} + \ldots + a_n x_{out}(t) =$$

$$b_0 \frac{d^m x_{in}(t)}{dt^m} + b_1 \frac{d^{m-1} x_{in}(t)}{dt^{m-1}} + \ldots + b_m x_{in}(t), \quad m < n.$$

One of the mathematical descriptions of the dynamic system is its transfer function *H*(s). The transfer function is the ratio of the Laplace transforms of the output and input processes of the system:

$$H(s) = \frac{x_{out}(s)}{x_{in}(s)},$$

where:

$$x(s) = \int_0^\infty x(t) e^{-st} dt.$$

Here $s = \sigma + i\omega$ is a complex frequency and $x(s)$ is the Laplace image of the function $x(t)$. The Laplace transform may be interpreted as an operator that transforms a function of time t to a function of complex frequency s. Being similar to the Fourier transform, the Laplace transform is a complex function of a complex variable s while the Fourier transform is a complex function of a real variable ω.

An important feature of the Laplace transform, which predetermined its wide applications in scientific and engineering fields, is that many relations and operations on the original functions correspond to simpler operations of their Laplace images. For instance, the convolution of two functions is reduced to a multiplication operation, and the linear differential equations become algebraic. The above differential equation in the space of Laplace images takes the following form:

$$\begin{aligned} &a_0 s^n x_{out}(s) + a_1 s^{n-1} x_{out}(s) \ldots + a_n x_{out}(s) \\ &= \left(a_0 s^n + a_1 s^{n-1} + \ldots + a_n\right) \cdot x_{out}(s) \\ &= \left(b_0 s^m + b_1 s^{m-1} + \ldots + b_m\right) \cdot x_{in}(s). \end{aligned}$$

Thus, the transfer function equals to the following expression:

$$H(s) = \frac{x_{out}(s)}{x_{in}(s)} = \frac{b_0 s^m + b_1 s^{m-1} + \ldots + b_m}{a_0 s^n + a_1 s^{n-1} + \ldots + a_n},$$

which is completely determined by the coefficients of the differential equation describing the system. If k linear systems are connected in series, then the transfer function $H_\Sigma(\omega)$ of this series is equal to the product of the transfer functions $H_j(\omega)$ of these systems:

$$H_\Sigma(s) = \prod_{j=1}^{k} H_j(s).$$

The expression for the *FRF* can be obtained from $H(s)$ by simple replacement of "s" with "$i\omega$":

$$H(\omega) = \frac{b_0 (i\omega)^m + b_1 (i\omega)^{m-1} + \ldots + b_m}{a_0 (i\omega)^n + a_1 (i\omega)^{n-1} + \ldots + a_n}.$$

As an example, consider the behavior of the *GTE* in steady-state mode under external influences. Recall that such a situation was discussed in section 3.3.3, where the engine as a dynamic system had an input in the form of fuel consumption $G_f(t)$, and its output was the rotor speed $n(t)$. The engine as a dynamic system may be described by the ordinary differential equation of the first order:

$$\tau_n \frac{dn(t)}{dt} + n(t) = \mathrm{K} \cdot G_f(t).$$

Here τ_n is a time constant which characterizes the inertial lag in the change of $n(t)$ when $G_f(t)$ is changed; K is a gain of the dynamic system in steady-state mode. Let the fuel supply be changed at the moment t_0 in a form of a step increase of its value. Such a change may be mathematically described by the Heaviside step function:

$$1(t-t_0)=\begin{cases}1, & t\geq t_0\\ 0, & t<t_0\end{cases}.$$

The corresponding engine behavior (change of the rotor speed) is illustrated in Fig. 4.1.

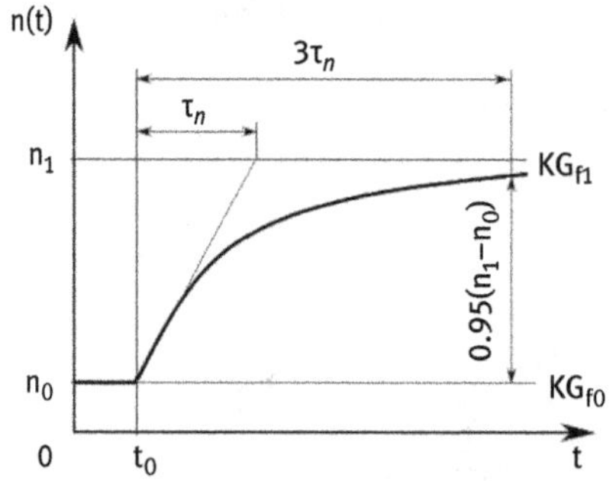

Fig. 4.1: Change in *n(t)* after increasing $G_f(t)$.

A value n_1 corresponds to a new steady-state mode when the transient processes in the engine are finished.

The appropriate *FRF* of the engine will be:

$$H(\omega)=\frac{K}{1+i\omega\tau_n}=\frac{K(1-i\omega\tau_n)}{(1+i\omega\tau_n)(1-i\omega\tau_n)}$$

$$=\frac{K}{1+\omega^2\tau_n^2}-i\cdot\frac{K\omega\tau_n}{1+\omega^2\tau_n^2}.$$

Correspondingly, the amplitude and phase components of the engine's *FRF* have the following forms:

$$A(\omega)=\frac{K}{\sqrt{1+\omega^2\tau_n^2}};$$

$$\Phi(\omega)=\tan^{-1}(-\omega\tau_n).$$

These components are usually represented in a form of the plots against the logarithmic frequency axes (Fig. 4.2).

The value $f=f_c=1/\tau_n$ is called the cutoff frequency; a corresponding value of $A(f_c)=-3\,dB$. After f_c, a roll-off rate of $A(f)$ is $-20\ dB$ per decade. The phase component of the *FRF* has a zero value almost up to a frequency of $0.02f_c$. Afterwards, it

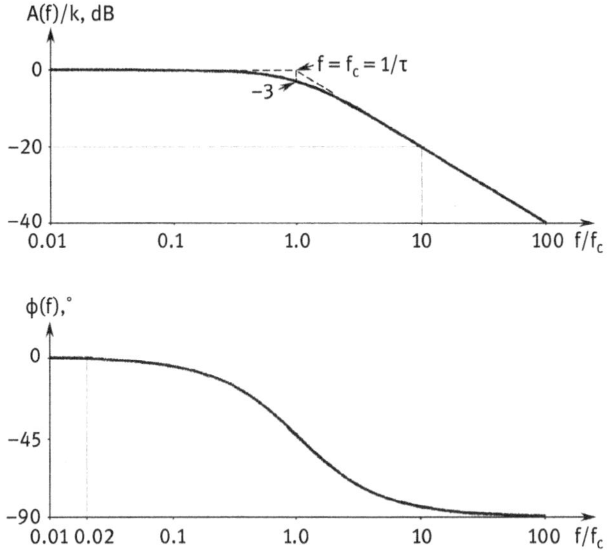

Fig. 4.2: The amplitude and phase plots.

changes to a value of $-45°$ at f_c and asymptotically tends to $-90°$. The frequency range $0...f_c$ is called bandwidth in which most of the energy of the system output is concentrated.

Modern *GTEs* have values of $\tau_n > 1\,sec$ for majority of operational modes, that is, the cutoff frequency is $f_c < 1\,Hz$. For the *IR* mode, the time constant is an order of magnitude greater ($f_c < 0.1\,Hz$). This is due to the fact that in the *IR* mode the engine has a low rate of fuel consumption. As such, a longer time lag is needed to accumulate energy in order to change the rotor speed.

In the *AMIS*, the process $X(t)$ is presented by its values at moments of time t_i. Such a conversion of a continuous process $X(t)$ into a set of samples $X(t_i)$ is carried out by the *ADC* described in section 2.1.

Sample $X(t_i)$ is a readout of $X(t)$ at a point in time $t_i = \Delta t \cdot i$. The interval Δt is referred to as the sampling interval. The theoretical base of the discrete representation of the process $X(t)$ is the Nyquist-Shannon-Kotelnikov theorem. Its other name is the sampling theorem.

Particularly, Shannon has formulated it in the following terms. If a process $X(t)$ contains no frequencies higher than f_m, then this process is completely determined by giving its ordinates at a series of points spaced $\Delta t = 1/(2f_m)$ apart. The value $f_N = 2f_m$ is called the Nyquist frequency (rate).

The theorem sets up a maximum value of interval Δt, however, for practical purposes its value is obtained from the expression of the form:

$$\Delta t = \frac{1}{\mathrm{k} \cdot f_N},$$

where $\mathrm{k} > 1$ is a margin coefficient whose value depends on how much a real process is approximated by a process featuring the spectrum limit f_m as well as on the purpose of subsequent data processing.

4.1.2 Random processes

In section 2.1, the measurement channel model was established in the following form:

$$\tilde{X} = X + \Xi.$$

Taking into account the time factor, one can write:

$$\tilde{X}(t) = X(t) + \Xi(t),$$

where $X(t)$ is the measured deterministic process and $\Xi(t)$ is the measurement (instrumental) noise.

By a random process, one implies a time function whose instantaneous values are random variables. Samples $\xi_1, \ldots, \xi_i, \ldots, \xi_n$ of the instrumental noise at discrete moments $t_1, \ldots, t_i, \ldots, t_n$ may be viewed as a random vector $\xi^{\mathrm{T}} = (\xi_1, \ldots, \xi_i, \ldots, \xi_n)$. Note that the order in this sequence is of importance for a random process whereas the multivariate vector mentioned in section 1.1.1 had indexes for notational convenience only.

The random series $\xi_1, \ldots, \xi_i, \ldots, \xi_n$ is referred to as stationary if the distribution function $F(\xi_{i+1}, \ldots, \xi_{i+n})$ of any of its n values is independent of index "i". In this case any n consecutive variables have the same distribution regardless of their position in the series with expectation

$$\mu_\xi = M[\xi_i] = \int_{-\infty}^{\infty} \xi_i f(\xi_i)\, d\xi_i,$$

and variance

$$D_\xi = \sigma_\xi^2 = M\left[\left(\xi_i - \mu_\xi\right)^2\right] = \int_{-\infty}^{\infty} (\xi_i - \mu_\xi)^2 f(\xi_i)\, d\xi_i.$$

The relationship between two values ξ_i and ξ_{i+k} of the stationary series with a lag k is characterized by the covariance:

$$\gamma_k = M\left[\left(\xi_i - \mu_\xi\right)\left(\xi_{i+k} - \mu_\xi\right)\right] = \gamma_{-k},$$

as well as by the correlation:

$$\rho_k = \frac{\gamma_k}{D_\xi} = \rho_{-k}.$$

Sequences of process covariance and correlation are termed the covariance and correlation functions, respectively.

The Fourier transformation of the covariance function of the stationary process is called a one-sided power spectral density:

$$S(\omega) = \frac{1}{\pi} \sum_{i=-\infty}^{\infty} \gamma_i \cos(\omega \cdot i), \quad \Delta t = 1, \; 0 \leq \omega \leq \pi.$$

The expression for covariance represented in terms of $S(\omega)$ is:

$$\gamma_k = \int_0^\pi \cos(\omega \cdot k) S(\omega) d\omega,$$

and variance, in particular, is:

$$D_\xi = \gamma_0 = \int_0^\pi S(\omega) d\omega.$$

The power spectral density, which called the spectrum going forward, shows the distribution of the random process variance (power) within a continuous frequency range $0 \ldots \pi$. Hence, the value $S(\omega) \cdot d\omega$ may be interpreted as an approximate portion of the process variance within the frequency range $\omega \ldots (\omega + d\omega)$. An example of the spectrum $S(\omega)$ of instrumental noise of some measurement channel is represented in Fig. 4.3.

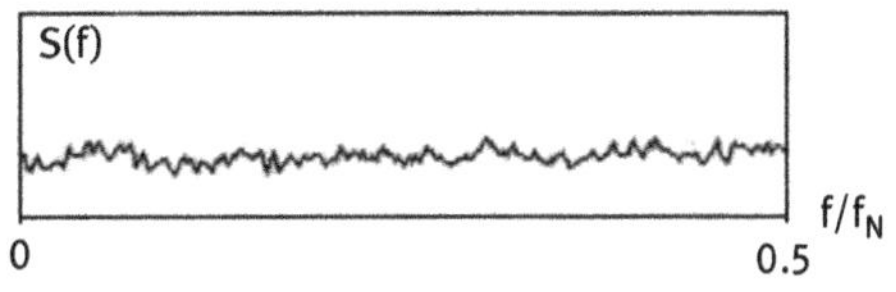

Fig. 4.3: Spectrum of the instrumental noise.

A random process $\Xi(t)$ with a constant value of $S(\omega) = D_\xi/\pi$ in the range $0 \ldots \pi$ is called "white" noise with a finite bandwidth. The covariance function of such a process is:

$$\gamma(\tau) = D_\xi \cdot \frac{\sin(\pi\tau)}{\pi\tau}.$$

In case of a discrete sequence $\xi_1, \ldots, \xi_i, \ldots, \xi_n$, the time lag $\tau_k = \Delta t \cdot k$ is equal to k with $\Delta t = 1$. Since for $k = 1, 2, \ldots$, the value of $sin(\pi k)/\pi k = 0$, there is only the value of $\gamma(\tau = 0) = \gamma_o = D_\xi$. In other words, samples of the instrumental noise $\Xi(t)$ are uncorrelated. The value of D_ξ can be estimated as $\hat{\sigma}_{\tilde{X}}^2$ obtained during the static calibration of the measurement channel discussed in section 3.1.2.

4.2 Dynamic calibration of measurement channel

After static calibration of the measurement channel, it can be considered as a linear dynamic system with unity steady-state gain [8]. It means that the inertial properties of the measurement channel can be fully described by its *IRF*. Therefore, the purpose of the dynamic calibration of the measurement channel is the determination of its weight function $h(t)$.

4.2.1 Calibration procedure implementation

The procedure for dynamic calibration as whole follows the scheme shown in Fig. 3.1. The only difference is that instead of the constant magnitude of the value X, the reference process $X(t)$ must be reproduced at the channel input. If $X(t)$ will be the Dirac's delta function $\delta(t)$, the channel output $\tilde{X}(t)$ will correspond to $h(t)$.

It should be noted that in practice, the delta function can be reproduced only in the form of some approximations. Therefore, there is a methodical problem to assess the conformity of the real reference process to the delta function. In [2], possible approximations of the reference process were considered in the form of a rectangle and an isosceles triangle. It was shown that the spectra of both approximations have conformity to the spectrum of the $\delta(t)$ only in a limited frequency region. In other words, $\tilde{X}(t)$ will correspond to the $h(t)$ with some kind of error. In addition to this methodological issue, the immediate (after the direct "step") reproduction of the reverse "step" is a complex technical task in itself. Therefore, from the standpoint of the technical implementation, it is preferable to use the Heaviside function $1(t)$ as the reference process. It should be noted that the methodical aspect of non-ideality of such a reference process is greatly simplified. In reality, a source of the reference process being a physical device will produce a certain transient process instead of an ideal $1(t)$. This means that this device has some transfer function $H_{ref}(s)$ of its own. Therefore, the calibration results will describe the properties of the calibration device and the *MC*. In other words, they will allow us to determine the common transfer functions:

$$H_\Sigma(s) = H_{ref}(s) \cdot H_{ch}(s).$$

With known $H_{ref}(s)$ and $H_\Sigma(s)$, the transfer function $H_{ch}(s)$ can be found from above expression. In practice, the source of the reference process in form of the function $1(t)$ can be characterized by only the time constant τ_{ref}. From the Fig. 4.2, it follows that if the cutoff frequency of the calibrated channel is less than $\frac{1}{50\tau_{ref}}$, then the influence of $H_{ref}(s)$ can be neglected.

As an example, Fig. 4.4 represents a scheme of a calibration device reproducing a pressure step [2]. This device can be used to provide a pressure step at the inlet of the pneumatic pipeline of the pressure measurement channel shown in section 2.1 (Fig. 2.4).

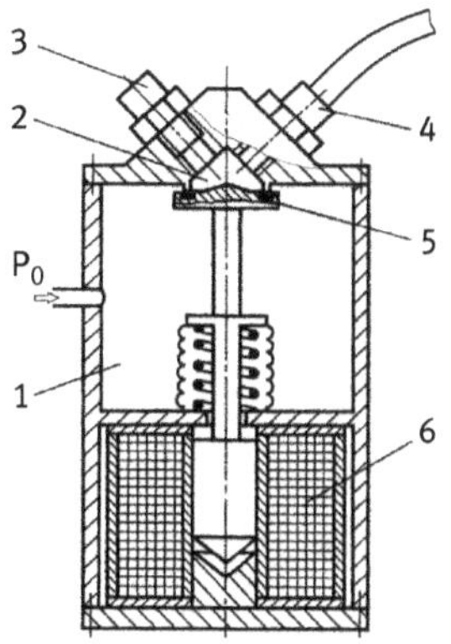

Fig. 4.4: A scheme of a calibration device.

The main components of the device are: 1 – charging cavity; 2 – working cavity; 3 – reference pressure sensor; 4 – pipeline inlet; 5 – high-speed valve; 6 – electromagnet (solenoid). The valve allows a quick bypass of compressed air from the charging cavity to the working cavity. Appropriate selection of the volume of these two cavities and an air pressure P_0 ensures the required steady-state value of the air pressure in the pipeline.

The time constant of this device is $\tau_{ref} \approx 0.008$ sec which corresponds to the cutoff frequency 125 Hz. Therefore, the inertia of the considered device can be neglected when it is used for the calibration of channels with cut-off frequencies ≤ 2.5 *Hz*.

A reference sensor having a time constant 20 times smaller than τ_{ref} is used to determine the moment t_o of occurrence of a pressure step in the working cavity.

As it is known, there is a one-to-one relationship between the Dirac's delta and Heaviside's step functions:

$$\delta(t) = \frac{d1(t)}{dt}.$$

If a response $\tilde{x}(t)$ of the channel to the input in the form of the $1(t)$ is available, then the weight function may be determined by simple differentiation of the process $\tilde{x}(t)$:

$$\tilde{h}(t) = \frac{d\tilde{x}(t)}{dt}.$$

Recall that the process $\tilde{x}(t)$ is the sum of the channel response to the reference process and the channel instrumental noise:

$$\tilde{x}(t) = x(t) + \xi(t).$$

A discrete (at moments $t_j = \Delta t \cdot j$) representation of the channel output has the following form:

$$\tilde{x}(t_j) = \tilde{x}_j = x_j + \xi_j, \; j = 0, 1, 2, \ldots$$

First of all, we consider the determination of $h(t)$ by differentiating the channel response when its input is the Heaviside function. In the case of a presence of discrete values x_j, their derivative can be calculated using its approximation by the first-order finite difference:

$$\frac{dx(t)}{dt}\Big|_{t=t_{j-1}} \approx \frac{x_j - x_{j-1}}{\Delta t}.$$

As a result, the estimations of the derivative of the process $x(t)$ at point t_{j-1} can be done by calculating the value $(x_j - x_{j-1})$ and dividing it by the sampling interval Δt. In section 4.1.1, it was noted that the sampling interval of the continuous process is related to the Nyquist frequency:

$$\Delta t = \frac{1}{f_N}.$$

From the definition of the first order derivative:

$$\frac{dx(t)}{dt}\Big|_{t=t_{j-1}} = \lim_{\Delta t \to 0} \frac{x(t_j) - x(t_j - \Delta t)}{\Delta t},$$

it follows that with decreasing the time interval Δt the first-order difference approaches the value of the derivative. This means that the smaller the sampling interval, the more accurate the estimate of the weight function.

Let's turn to the second term of $\tilde{x}(t)$ – the instrumental noise $\xi(t)$. Its derivative can be represented in the following form:

$$\frac{d\xi(t)}{dt}\Big|_{t=t_{j-1}} \approx \zeta_{j-1} = \frac{\xi_j - \xi_{j-1}}{\Delta t}.$$

Therefore, ζ_{j-1} is a random variable with *rms* value related to the *rms* error of a measurement channel $\hat{\sigma}_{\tilde{X}}$:

$$\hat{\sigma}_{\zeta} = \frac{\sqrt{2}}{\Delta t} \cdot \hat{\sigma}_{\tilde{X}}.$$

Obviously, the smaller the sampling interval Δt, the greater the statistical error of estimates of $h(t)$.

These results demonstrate the existence of two errors of estimates of the channel weight function. One of them, being deterministic (bias), is caused by calculating the derivative via the first-order difference. Another error, being random, is due to the instrumental noise inherent to the *MC*. Both of them are influenced by a change in the value of sampling interval in opposite ways. The analysis of the behavior of these errors as a function of Δt can be helpful to make a choice of sampling interval for a particular measurement channel.

Consider as an example a channel corresponding to the first-order differential equation. This means that the channel can be modeled by a low-pass filter with a cutoff frequency $f_c = 1/\tau_{ch}$. This type of description of a dynamic measurement channel is often used in practice. The weight function of the low-pass filter corresponds to the following expression:

$$h(t) = \frac{1}{\tau} e^{-t/\tau_{ch}}.$$

If the input of this channel is affected by the function $1(t)$, then the output (transient) process of the measurement channel is described by the following expression:

$$x(t) = \left(1 - e^{-t/\tau_{ch}}\right).$$

The measuring scale of the channel in the form $0 \ldots 1$ is adopted to simplify the following discussions of precision issues.

First of all, let's consider the systematic error of estimates of the weight function caused by calculating the first difference of the transient process (Fig. 4.5).

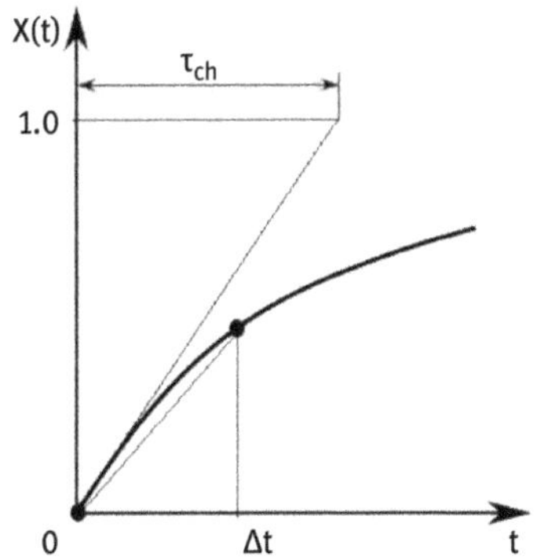

Fig. 4.5: Transient process in the measurement channel.

The sampling interval Δt can be associated with the time constant of the low-pass filter in the following way:

$$\Delta t = \frac{\tau_{ch}}{\mathrm{k}}.$$

The theoretical value of the weight function at moment $t = 0$ is determined as:

$$h(t=0) = h_0 = \frac{dx(t)}{dt} = \frac{1}{\tau_{ch}}.$$

On the other hand, the estimate of h_0, calculated through the first-order difference of the series x_i, can be represented as:

$$\frac{dx(t)}{dt}\Big|_{t=0} \approx \frac{x(\Delta t)}{\Delta t} = \frac{\mathrm{k}}{\tau_{ch}} \cdot \left(1 - e^{-\Delta t/\tau_{ch}}\right)$$

$$= \frac{\mathrm{k}}{\tau_{ch}} \cdot \left(1 - e^{-1/\mathrm{k}}\right).$$

Thus, the relative bias δ_h of h_0 estimates can be defined by the following expression:

$$\delta_h = \frac{\frac{dx(t)}{dt}\Big|_{t=0} - h_0}{h_0} = \frac{\frac{\mathrm{k}}{\tau_{ch}} \cdot \left(1 - e^{-\frac{1}{\mathrm{k}}}\right)}{\frac{1}{\tau_{ch}}} - 1$$

$$= \mathrm{k} \cdot \left(1 - e^{-\frac{1}{\mathrm{k}}}\right) - 1.$$

The value of $\delta_h < 0$ because for $\mathrm{k} = 1$ the value of $\delta_h \approx -0.37$, that is, the value of h_j is always underestimated. It should be noted that the rate of the δ_h change decreases when k increases.

As demonstrated earlier, the differentiation of the instrumental noise ξ_j leads to a statistical error whose relative value can be determined as:

$$\delta_\zeta = \frac{\hat{\sigma}_\zeta}{h_0} = \sqrt{2} \cdot \mathrm{k} \cdot \hat{\sigma}_{\tilde{X}}.$$

This error may be characterized by a confidence interval $\Delta_\zeta(P = 0.95)$ as follows:

$$\Delta\zeta = \sqrt{2} \cdot \mathrm{k} \cdot \hat{\Delta}_{\tilde{X}}.$$

where $\hat{\Delta}_{\tilde{X}}$ is the confidential interval of the measurement channel error. The linear dependence of Δ_ζ on k implies a constant rate of change.

Finally, the total error of differentiating the measurements $\tilde{x}_j$ can be described by the following expression:

$$\delta_\Sigma = |\delta_h| + \Delta_\zeta = 1 - \mathrm{k}\left(1 - e^{-\frac{1}{\mathrm{k}}}\right) + \sqrt{2} \cdot \mathrm{k} \cdot \hat{\Delta}_{\tilde{X}}.$$

Recall that with increasing the value of k, the behavior of the systematic (bias) and random components of δ_Σ is opposite. In particular, the value of $|\delta_h|$ decreases with growing k while the value of Δ_ζ increases. Furthermore, the rate of changing $|\delta_h|$ and Δ_ζ is different as well, and for small values of k, the bias dominates. Due to the fact that the rate of bias change falls with increasing k and the rate of Δ_ζ is constant, the values of these errors will be equal for a certain k. Such an equality of these errors can be written in the following form:

$$1-\mathrm{k}\left(1-e^{-\frac{1}{\mathrm{k}}}\right)=\sqrt{2}\cdot\mathrm{k}\cdot\hat{\Delta}_{\tilde{X}},$$

that is, it depends only on a value of $\hat{\Delta}_{\tilde{X}}$. In particular, for the value $\hat{\Delta}_{\tilde{X}}=0.003(0.3\%FS)$, the estimate of the bias of h_0 compares with the statistical error at $\mathrm{k}=11(|\delta_h|=0.044$ and $\Delta_\zeta=0.047)$ as it is shown in Fig. 4.6.

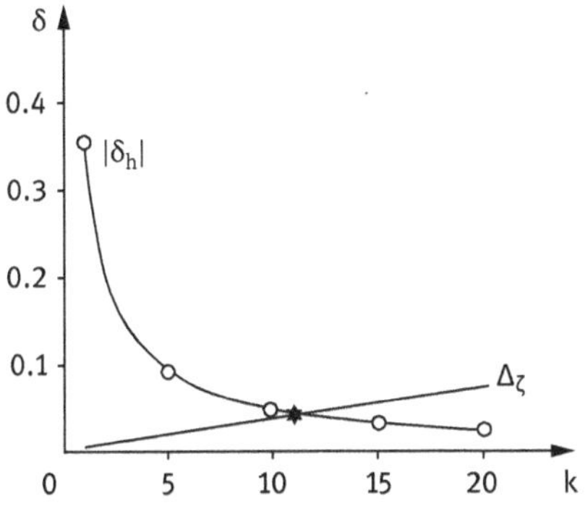

Fig. 4.6: Changing values of $|\delta_h|$ and Δ_ζ.

Additionally, at this point the negative rate of the value $|\delta_h|$ reduces to the constant positive rate of the statistical error. This means that the first derivative of δ_Σ is zero indicating the achievement of the optimum. With subsequent increases of the value of k the positive rate of the Δ_ζ begins to dominate because the rate of $|\delta_h|$ continues to decline. Consequently, the total error δ_Σ acquires a growing positive rate. This means that its second derivative is positive, that is, the optimum is the minimum. In other words, at this point the error of estimates of h_0 has a minimal value.

It should be noted that an increase of δ_Σ values will occur with a much lower rate than its decrease before the minimal value. For example, when $\mathrm{k}=20$, the value $\delta_\Sigma=0.11$, that is, it will increase only by 1.2 times. Note that when value k is changed from 1 to 10, the value of δ_Σ decreased by ≈ 4 times (from 0.372 to 0.091).

In addition to taking into account the purpose of processing calibration data, the choice of k must ensure the Nyquist rate as well. For an adequate discrete representation of the output process of the low-pass filter, it is necessary to know its maximum frequency f_m. From Fig. 4.2 it follows that when $f=10f_c$, the value of the amplitude component $A(f)=0.1$. In other words, beyond this frequency the harmonics of the input process $x(t)$ are attenuated by more that an order of magnitude. If the value $f_m = 10f_c$ will be assumed as the frequency bandwidth (passband), the Nyquist frequency will be equal to:

$$f_N = 2f_m = \frac{20}{\tau_{ch}}.$$

This means that the interval Δt will be tied with the constant time of the channel via the following expression:

$$\Delta t = \frac{1}{f_N} = \frac{\tau_{ch}}{20}.$$

If we turn to the above example with $\hat{\Delta}_{\tilde{X}} = 0.003$, the value k = 20 leads to follows errors of h_0 estimates:

$$\delta_\Sigma = |\delta_h| + \Delta_\zeta \approx 0.025 + 0.085 = 0.11.$$

This example quantitatively illustrates the effect of numerically differentiating the measurements $\tilde{x}_j$ of the transient process $x(t)$. One can see that the instrumental error 0.3% (*FS*) has led to an increase in the error of estimates of h_0 by a factor of 28 (up to 8.5%). It should be noted that the random error dominates exceeding the bias by 3.4 times.

At the moment $t = 0$, the value of $|\delta_h|$ is 0.025. At the points $t = \tau_{ch}$, $2\tau_{ch}$, $3\tau_{ch}$, the values of biases related to $h_0 = 1/\tau_{ch}$ are equal to 0.009, 0.003, 0.001. This means that they are less than the random error by more than an order of magnitude. Due to the dominance of Δ_ζ, a possible decrease of the error δ_Σ is mainly associated with a decrease in the magnitude of Δ_ζ. To do this, it is needed to obtain l realizations of $\tilde{x}(t)$ as their averaging will decrease the value of Δ_ζ by a factor of $\sqrt{l}$. In addition to the multiple calibrations, another approach to reduce the influence of the random noise ζ_j will be demonstrated in the following section.

Estimates $\tilde{h}_j$ obtained by numeric differentiation of the series $\tilde{x}_j$ will contain all information about inertial properties of the measurement channel.

4.2.2 Approximation of frequency response

The estimates $\tilde{h}_j (j = 1, \ldots, n)$ obtained during the channel calibration can be used to calculate the *FRF* as follows:

$$\tilde{H}(\omega) = \sum_{j=1}^{n} \tilde{h}_j \cdot e^{-i\omega} = \tilde{U}(\omega) + i\tilde{V}(\omega).$$

The use of the *FRF* is preferable, since it allows one to "compress" the amount of experimental data obtained in the time domain. The point is that the bandwidth of measurement channels is usually an order of magnitude larger that the frequency spectrum of the processes observed in the test object. Therefore, only values of $\tilde{H}(\omega)$ in the frequency spectrum inherent to the test object need to be taken into account. Moreover, the parametric (compact) description of channel inertial properties may be done in the form of a transfer function. There is sufficient presentation [9] of methods for the identification of transfer functions of dynamic systems using experimental data. Some of them are already implemented in *COTS* products.

However, for illustrative purpose consider the method presented in [2]. According to this method, the evaluation of factors $a_q(q=1,n)$ and $b_r(r=1,m)$ of the *TF* occurs by minimizing a functional *I* having the following form:

$$I=\sum_{j=1}^{k}\left|\tilde{H}(\omega_j)-\frac{B_m(\omega_j)}{A_n(\omega_j)}\right|^2$$

$$=\sum_{j=1}^{k}\left|\tilde{H}^{(\omega_j)}-\frac{1+b_1(i\omega_j)+\ldots+b_m(i\omega_j)^m}{1+a_1(i\omega_j)+\ldots+a_n(i\omega_j)^n}\right|^2\underset{a_q,b_r}{\Rightarrow}\min.$$

An effective solution to this task requires setting fairly close initial values of a_q and b_r, which often presents significant complexity. To avoid this, the initial task was reduced to an iterative procedure to minimize the quadratic function of factors a_q and b_r. The lth step of that procedure is formalized in the following form:

$$I_l=\sum_{j=1}^{k}\alpha_{j,l-1}^2\left|\tilde{H}(\omega_j)\cdot A_n(\omega_j)-B_m(\omega_j)\right|^2$$

$$=\sum_{j=1}^{k}\alpha_{j,l-1}^2|\beta_j|^2\underset{a_q,b_r}{\Rightarrow}\min,$$

Here $\alpha_{j,l-1}^2=1/|A_n^{(l-1)}(\omega_j)|^2$ is calculated using the values of factors a_q obtained in the previous $(l-1)^{th}$ step.

Conditions of the I_l minimum determined as

$$\begin{aligned}&\sum_{j=1}^{k}\alpha_{j,l-1}^2\frac{\partial|\beta_j|^2}{\partial a_q}=0;\\&\sum_{j=1}^{k}\alpha_{j,l-1}^2\frac{\partial|\beta_j|^2}{\partial b_r}=0\end{aligned}\qquad(8)$$

lead to a system of $(n+m)$ equations with respect to a_q and b_r. These equations are linear because the following are linear expressions with respect to a_q and b_r:

$$\frac{\partial|\beta_j|^2}{\partial a_q} = [\tilde{H}(\omega_j)\tilde{H}^*(\omega_j)A_n(\omega_j) + (-1)^q\tilde{H}(\omega_j)\tilde{H}^*(\omega_j)A_n(\omega_j)$$

$$-\tilde{H}(\omega_j)B_m(-\omega_j) - (-1)^q\tilde{H}^*(\omega_j)B_m(\omega_j)](i\omega_j)^q;$$

$$\frac{\partial|\beta_j|^2}{\partial b_r} = [\tilde{H}^*(\omega_j)A_n(-\omega_j) + (-1)^r\tilde{H}(\omega_j)A_n(\omega_j)$$

$$-B_m(-\omega_j) - (-1)^r\tilde{H}^*B_m(\omega_j)](i\omega_j)^r.$$

Thus, the approximation procedure is reduced to solving the system of $(n+m)$ linear algebraic equations on its lth iteration step.

The initial values of $\alpha_{j,0}^2$ are set on *a priori* information about n, m and values of a_q. The presence of this data accelerates the convergence of the procedure. If there is no such data, then the order of the *TF* can be sequentially modified from $n = 1$ and $m = 0$ to $n = 2$, $m = 1$ and so on. It should be noted that the convergence of the procedure remains quite effective even with an arbitrary choice of a_q, for instance, $a_q = 0$.

The iterative process of approximating the *FRF* ends when two conditions are satisfied:

$$|I_l - I_{l-1}| \le \delta_0;$$

$$\sum_{q=1}^{n}\left|a_q^{(l)} - a_q^{(l-1)}\right| + \sum_{r=1}^{m}\left|b_r^{(l)} - b_r^{(l-1)}\right| \le \delta_1.$$

Here δ_0 and δ_1 are given numbers characterizing the convergence of the algorithm and accuracy of estimates, respectively.

This approach was tested using analog models of dynamic systems with transfer functions up to $n \le 4$. During testing, a high rate of convergence of solutions was demonstrated: usually, a stabilization of factor estimates in the third significant digit occurred with $l \le 5$. When the normal measurement noise with $\sigma_{ch} = 10\%(FS)$ was added to a single realization of model transient processes, the relative error in the factor estimates did not exceed 2%. Hence, the approximation procedure reduces the influence of the instrumental noise in addition to the averaging of several realizations of the transient process mentioned in section 4.1.1.

Implementation of this method may be illustrated by the example of determining the transfer function of the channel for measuring the thrust of the *GTE*. The dynamic platform of the thrust measurement system (*TMS*) with an engine was loaded in the direction of flight by means of a steel cable with known tensile strength (Fig. 4.7).

The platform was preloaded to F_{R_1} to ensure the observations of a possible "overshoot" in the limits of a channel measurement scale. When the tensile strength (F_{R_2}) was reached, the cable broke, and the platform began free movement to F_{R_1} (Fig. 4.8).

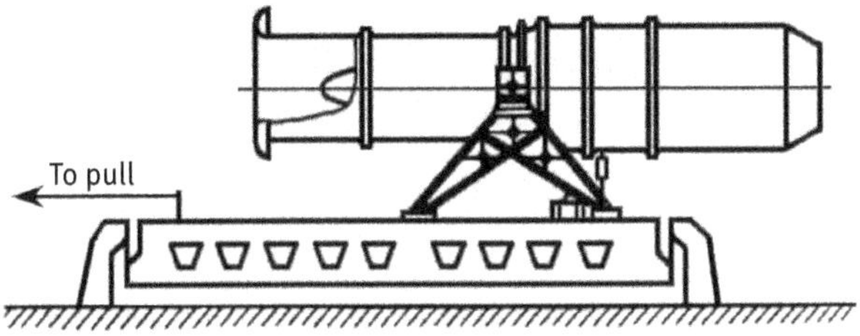

Fig. 4.7: Dynamic calibration of the *TMS*.

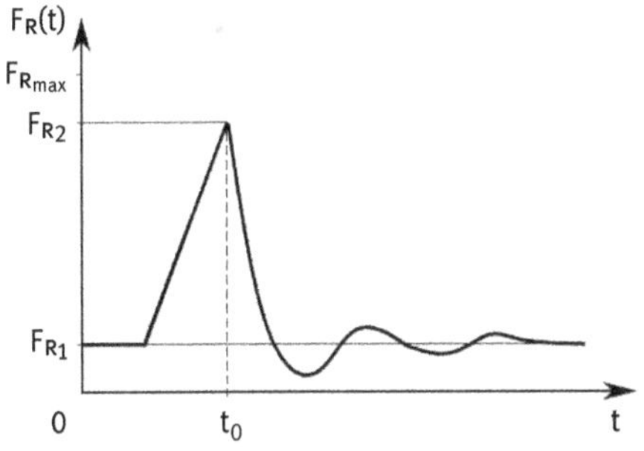

Fig. 4.8: The *TMS*' transient process.

The steel cable was included in the electrical circuit of the platform motion detector. At the moment of its rupture (t_0), the electrical current through the cable was stopped, and the detector issued a command to start measuring the transient process $F_R(t)$. This procedure was repeated several times. Averaging measurement results made it possible to reduce the influence of the instrumental noise on the calibration data. Subsequent processing of these data allowed to determine the estimates of the *FRF* presented in Fig. 4.9.

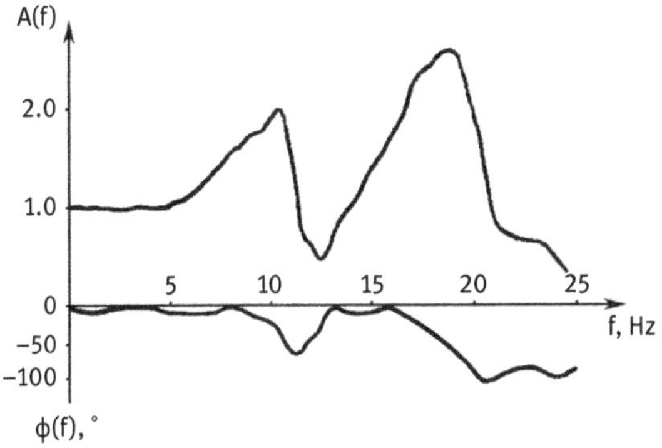

Fig. 4.9: The experimental *FRF* of the *TMS*.

As you can see, the frequency response contains two resonant frequencies in the regions of 11 *Hz* and 18 *Hz*. They are related to the flexure plates (Fig. 4.10) through which the dynamic platform is connected to the frame of the *TMS*.

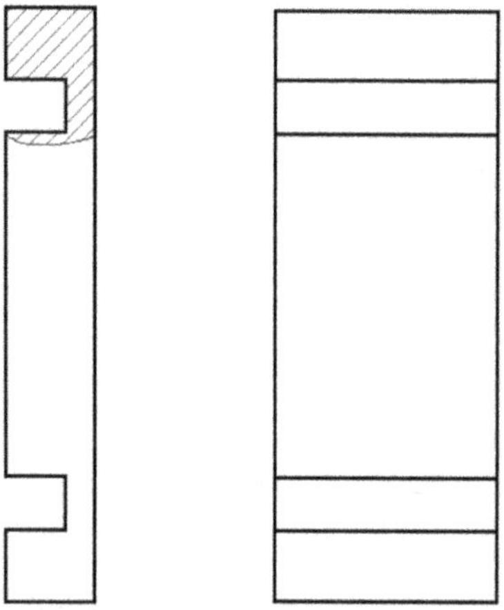

Fig. 4.10: The flexure plate design.

The oscillations arising from the free movement of the platform depend only on properties of the material, shape and size of the plates as well as the design of their fastening. Observed frequencies are associated with longitudinal-torsional oscillations of the flexure plates. The torsional frequencies can occur if the center of mass of "platform + engine" is not correlated with the point of the cable attachment. An additional reason for their appearance is the direction of the cable tension that is not strictly parallel to the longitudinal axis of the planform.

The design of the *TMS* allows the movement of the planform only in the longitudinal direction corresponding to the direction of the engine thrust vector. As such, the damping of the 11 *Hz* frequency is less than the damping at the 18 *Hz* frequency.

Earlier, it was noted that the time constant of the *GTE* is measured in seconds and for the *IR* mode, it can be an order of magnitude larger. Besides, the change of the engine modes is carried out under the control of the *ECU*. This unit implements dedicated algorithms that ensure stable, efficient and safe engine operation. In addition, their implementation leads to a smooth change in engine parameters. As such, the approximate frequency width of the *FRF* can be limited to a range of $0 \ldots 2\,Hz$. The approximating *TF* of the thrust measurement channel was defined with $n = 2$ and $m = 1$.

The values of $\tilde{H}(\omega)$ in the above frequency range were approximated by utilizing the system (8). The following view of the transfer function was yielded:

$$\tilde{H}(s) = \frac{1.0 + 0.116s}{1.0 + 0.124s + 0.0031s^2}.$$

The phase component of the corresponding *FRF* is represented in Fig. 4.11.

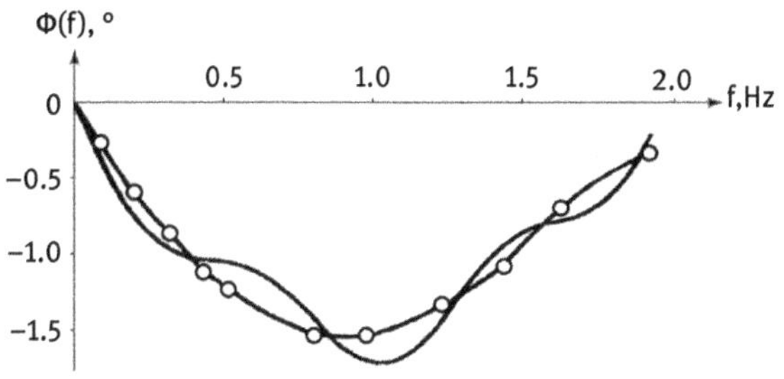

Fig. 4.11: Phase component of the *FRF*.

Here, the unbroken line represents the calibration data, and the line marked by circles corresponds to the approximating frequency response. It can be seen that there is a good match of calibration and approximation data. The term "approximation" is used instead of identification, since the obtained *TF* corresponds only to a part of the frequency width of the measurement channel.

4.3 Estimation of dynamic characteristics

As it was mentioned earlier, the inertial properties of a measurement channel affect the input process $x_{in}(t)$. The corresponded transformation into the output process $x_{out}(t)$ is determined by the weight function $h(t)$ of the *MC*. The result of such a transformation is quantitatively described by the convolution equation which relates to the Fredholm integral equation of the first kind:

$$\int_0^t h(t,\tau)x_{in}(\tau)d\tau = x_{out}(t).$$

Therefore, the knowledge of $h(t)$ and $x_{out}(t)$ gives a theoretical possibility to mitigate the influence of channel inertia. Such a procedure to recover the measured process can be fulfilled by solving the above integral equation for $x_{in}(t)$.

In section 4.2.1, it was shown that the weight function $h(t)$ as well as input and output processes of the channel can be provided in discrete form:

$$h(t_j) = h(\Delta t \cdot j) = h_j,\ j = 0,\ (m-1);$$

$$x_{in}(t_i) = x_{in}(\Delta t \cdot i) = x_{in,\,i};$$

$$x_{out}(t_i) = x_{out}(\Delta t \cdot i) = x_{out,\,i},\ i = 0,\ (n-1).$$

If $\Delta t = 1$, then the convolution integral for time $t_k = \Delta t \cdot k$ can be replaced by a sum using, for instance, the formula of rectangles:

$$\sum_{j=0}^{m-1} h_{k-j} \cdot x_{in,k} = x_{out,k}$$

The presence of measurement results in the form of a vector $\boldsymbol{X}_{out}{}^T = (x_{out,0}, \ldots, x_{out,n-1})$ allows to compose a system of linear algebraic equations. In the operator form, it can be written as follows:

$$\mathbf{A}_h \cdot \boldsymbol{X}_{in} = \boldsymbol{X}_{out}. \tag{9}$$

Here, $\mathbf{A}_h$ is a matrix approximating the function $h(t, \tau)$, $\boldsymbol{X}_{out}$ is the vector of the source data, and $\boldsymbol{X}_{in}$ is the vector of the system solution. Thus, recovering the input process of the measurement channel is reduced to solving the system (9) with respect to the vector $\boldsymbol{X}_{in}{}^T = (x_{in,0}, \ \ldots, \ x_{in,n-1})$.

4.3.1 Tikhonov's regularization method

A French mathematician Jacques Hadamard has suggested to classify the mathematical task as correct (well-posed), if:

1) its solution exists;
2) this solution is unique;
3) the solution is stable, that is, small deviations in the source data lead to small deviations in the solution.

Formally, solving the system (9) requires an inversion of matrix $\mathbf{A}_h$. Bear in mind that the ij^{th} element of the inverse matrix $\mathbf{A}_h{}^{-1}$ equals to the algebraic complement A_{ji} of the element ji^{th} of the matrix $\mathbf{A}_h$ divided by its determinant $|A_h|$. In principle, the matrix $\mathbf{A}_h$ may be not well-conditioned, that is, its properties can approach the properties of a singular matrix whose determinant is zero. Poorly conditioned matrices can lead to arbitrarily large changes in the solution $\mathbf{X}_{in}$ for small deviations in $\boldsymbol{X}_{out}$ due to division by a determinant close to zero. When considering the measurement channel, m < n, that is, the square matrix $\mathbf{A}_h$ will have zero columns. This fact means that its determinant is zero. In other words, the third Hadamard's condition may not to be satisfied.

Moreover, the weight function is determined by the results of dynamic calibration with some error δ_Σ as it was shown in section 4.2.1. In addition, the source data $\boldsymbol{X}_{out}$, which are the results of measurements, are accompanied by instrumental noise. Utilising such data can lead to composing the system of incompatible equations. It means that the first Hadamard's condition is also not satisfied. These considerations allow us to classify the recovery task as an ill-posed task.

To solve ill-posed tasks, the Tikhonov's regularization method [10] is widely used. Russian mathematician A.N. Tikhonov introduced the concept of an approximate solution, which is sought through a regularizing algorithm (operator) as a way to solve ill-posed tasks.

The basis of the regularization method is the Tikhonov's theorem which can be represented as follows. Suppose that the matrix $\mathbf{A}_h$ and vector $\boldsymbol{X}_{out}$ satisfy the conditions ensuring the compatibility of the system (9), and $\boldsymbol{X}_{in}{}^0$ is the solution of this system. Moreover, experimental estimates $\widetilde{\mathbf{A}}_h$ and $\widetilde{\boldsymbol{X}}_{out}$ with an error δ are available.

Assume that there are increasing functions $\varepsilon(\delta)$ and $\alpha(\delta)$ tending to zero as $\delta \rightarrow +0$ and such that $\delta^2 \leq \varepsilon(\delta) \cdot \alpha(\delta)$. Then for any $\varepsilon > 0$, there is a positive number δ_0 such that for values $\delta < \delta_0$ and $1/\varepsilon(\delta) \leq \alpha \leq \alpha(\delta)$, the vector $\boldsymbol{X}_{in}{}^\alpha$ delivering the minimum to the functional

$$\Psi\left(\boldsymbol{X}_{in}, \widetilde{\mathbf{A}}_h, \widetilde{\boldsymbol{X}}_{out}\right) = \left\|\widetilde{\mathbf{A}}_h \cdot \boldsymbol{X}_{in} - \widetilde{\boldsymbol{X}}_{out}\right\| + \alpha\|\boldsymbol{X}_{in}\|^2,$$

satisfies the inequality $\|\boldsymbol{X}_{in}^0 - \boldsymbol{X}_{in}^\alpha\| \leq \varepsilon$.

In a generally understood interpretation, the essence of the Tikhonov's theorem can be interpreted as follows: under certain conditions, there is an approximate solution to the ill-posed task with an accuracy corresponding to the accuracy of estimates of $\widetilde{\mathbf{A}}_h$ and $\widetilde{\boldsymbol{X}}_{out}$ elements.

Functional Ψ is called a regularizing operator (algorithm) and it must ensure that if $\boldsymbol{X}_{in}{}^\alpha$ approaches the exact solution $\boldsymbol{X}_{in}{}^0$ as $\delta \rightarrow 0$, the distance between these vectors decreases.

The term $\alpha\|\boldsymbol{X}_{in}\|^2$ is called the task stabilizer. It is strongly convex due to the fact that it is quadratic. Therefore, the functional Ψ is also strongly convex, that is, it reaches its minimum value at a single point. In fact, the ill-posed task is replaced by the correct minimization task:

$$\Psi\left(\boldsymbol{X}_{in}, \widetilde{\mathbf{A}}_h, \widetilde{\boldsymbol{X}}_{out}\right) \underset{X_n^\alpha}{\Rightarrow} \min.$$

Therefore, for small enough values of $\alpha > 0$, the approximate solution $\boldsymbol{X}_{in}{}^\alpha$ can approach the exact solution $\boldsymbol{X}_{in}{}^0$ with a given accuracy. In other words, while it is not possible to actually solve the task, it is possible to get its approximate solution with any desired accuracy.

The search for an approximate solution of system (9) includes forming the operator Ψ and determining the parameter α using additional information about the task. Such *a priori* data is physically justified information about the properties of the test object, such as knowledge about the smoothness of the process $X_{in}(t)$. The work processes of test objects (machines and mechanical devices) are in accordance with the basic physical laws (conservation of energy and momentum, continuity of flows of work fluids, and so on). Therefore, *a priori* information can be a statement that neither the $X_{in}(t)$ nor its derivatives can be discontinuous functions. The choice of the regularization parameter α is very often consistent with accuracy of experimental data. Such additional data can significantly narrow the range of feasible solutions simplifying the search for $X_{in}{}^{\alpha}(t)$.

4.3.2 Recovery of transient processes

The regularization method has proven itself in solving a variety of ill-posed problems, including the recovery of measured processes. One approach proposed in [2] was implemented for recovery of transient processes. According to this, the search of the approximate solution $x_{in}^{\alpha}(t)$ is formulated as a minimization of the following regularizing operator:

$$\Psi\left[x_{in}^{\alpha}(t)\right] = \int_{-\infty}^{\infty} \left| \int_{0}^{t} h(t-\tau) \cdot x_{in}^{\alpha}(\tau) d\tau - \widetilde{x}_{out}(t) \right|^{2} dt$$
$$+ \int_{-\infty}^{\infty} \left\{ B_0 \cdot [x_{in}^{\alpha}(t)]^2 + \sum_{l=1}^{p} B_l \cdot \left[\frac{d^l x_{in}^{\alpha}(t)}{dt^l} \right]^2 \right\} dt.$$

The task stabilizer $\alpha\|\boldsymbol{X}_{in}\|^2$ in the following form of essentially characterizes the smoothness of the process $x_{in}(t)$:

$$\alpha \int_{-\infty}^{\infty} \left\{ B_0 \cdot [x_{in}^{\alpha}(t)]^2 + \sum_{l=1}^{p} B_l \cdot \left[\frac{d^l x_{in}^{\alpha}(t)}{dt^l} \right]^2 \right\} dt.$$

The concerned approach utilizes the *FRF* of a measurement channel, that is, the functional Ψ is minimized in the frequency domain. Its Fourier image has the following form:

$$\Psi\left[X_{in}^{\alpha}(\omega)\right] = \int_{-\infty}^{\infty} |H(\omega) X_{in}^{\alpha}(\omega) - \widetilde{X}_{out}(\omega)|^2 d\omega$$
$$+\alpha \int_{-\infty}^{\infty} \sum_{l=0}^{p} B_l \cdot \omega^{2l} \cdot |X_{in}^{\alpha}(\omega)|^2 d\omega,$$

where $X_{in}^{\alpha}(\omega)$ and $\widetilde{X}_{out}(\omega)$ are Fourier transforms of $x_{in}^{\alpha}(t)$, and $\widetilde{x}_{out}(t)$, accordingly.

The solution $X_{in}^{\alpha}(\omega)$ providing a minimum of the functional $\Psi\left[X_{in}^{\alpha}(\omega)\right]$ is determined by the extremum condition:

$$\frac{\partial\Psi\left[X_{in}^{\alpha}(\omega)\right]}{\partial X_{in}^{\alpha}(\omega)}=0.$$

From its fulfillment, the following type of approximate solution is found:

$$X_{in}^{\alpha}(\omega)=\frac{H^{*}(\omega)\cdot\widetilde{X}_{out}(\omega)}{A^{2}(\omega)+\alpha\cdot B_{p}(\omega^{2})}. \tag{10}$$

Here $A(\omega)$ is the amplitude component of the channel's *FRF*, and index "*" is a sign of the conjugate, and $B_p(\omega^2)=\sum_{l=0}^{p}B_l\omega^{2l}$. From the expression (10) it follows that the approximate solution $X_{in}^{\alpha}(\omega)$ is obtained from the source data $\widetilde{X}_{out}(\omega)$ using a recovery filter with the *FRF* of the form:

$$H_{\alpha}(\omega)=\frac{H^{*}(\omega)}{A^{2}(\omega)+\alpha\cdot B_{p}(\omega^{2})}.$$

It should be noted that in addition to the recovery function, this filter will suppress the instrumental noise that accompanies the source data $\widetilde{X}_{out}(\omega)$. This statement is justified by the fact that $H_{\alpha}(\omega)$ correlates with the frequency response $H(\omega)$ of the measurement channel.

Substitution of the $X_{in}^{\alpha}(\omega)$ into the functional $\Psi\left[X_{in}^{\alpha}(\omega)\right]$ transforms the latter to a function of the parameter α:

$$\Psi(\alpha)=\int_{-\infty}^{\infty}\frac{\alpha\cdot B_{p}(\omega^{2})\cdot|\widetilde{X}_{out}(\omega)|^{2}}{A^{2}(\omega)+\alpha\cdot B_{p}(\omega^{2})}d\omega.$$

It is clear that the recovery task is reduced to searching for a minimum of the function of only one variable α. This essentially simplifies the optimization procedure.

The procedure for finding the optimal value of α can be realized as a numerical search (computational trials) of the minimum of the function $\Psi(\alpha)$. The determined value α_{opt} is used to calculate $X_{in}^{\alpha}(\omega)$. The approximating solution $x_{in}^{\alpha}(t)$ will be the reverse Fourier transform of the obtained $X_{in}^{\alpha}(\omega)$.

Computational trials require the selection of the area of possible values of α. The upper limit of its value can be found from the condition that $H_{\alpha}(\omega)$ of the recovering filter must have a passband greater than the bandwidth of the *MC*. In other words, the filter should not affect the process $X_{in}(t)$ being recovered. Since the channel has an amplitude frequency response $A(\omega)$ that decreases monotonically as ω increases, this condition can be written as:

$$A_{\alpha}(\omega)=\frac{A(\omega)}{A^{2}(\omega)+\alpha\cdot B_{p}(\omega^{2})}\geq A(\omega).$$

It follows that:

$$A^2(\omega)+\alpha\cdot B_p\left(\omega^2\right)\leq 1.$$

From this inequality, the upper limit of the possible values of α is found as:

$$\alpha\leq\frac{1-A^2(\omega)}{B_0+\sum_{l=1}^{p}B/\omega^{2l}}. \tag{11}$$

Let the measurement channel have the cut-off frequency $\omega_c=2\pi/\tau_{\mathrm{ch}}$. The largest gain and bandwidth has an *AFR* of a low-pass filter due to its *AFR* roll-off with a rate of only −20 dB per decade. Hence, being the upper boundary of $A^2(\omega)$, the following expression can be used:

$$A^2(\omega)=\frac{1}{1+\tau_{ch}^2\omega^2}.$$

Substitution of this expression in the above inequality (11) leads to the following expression:

$$\alpha\leq\frac{\tau_{ch}^2\omega^2}{\left(1+\tau_{ch}^2\omega^2\right)\left(B_0+B_1\omega^2+\sum_{l=2}^{p}B_l\omega^{2l}\right)}.$$

Decreasing the denominator by $\left(1+\tau_{ch}^2\omega^2\right)$, B_0, and $\sum\limits_{l=2}^{p}B_l\omega^{2l}$ allows to determine the range of exercised values of α as:

$$0<\alpha<\frac{\tau_{ch}^2}{B_1}. \tag{12}$$

The reconstruction of the time domain process $x_{in}^{\alpha}(t)$ has its own specifics. The transient process shows a change in the values of $x(t)$ during transition from one steady-state to another, that is, its initial and final values are different. The Fourier transform of the measurement data $\widetilde{x}_{out}(t)$ assumes a periodic nature of this fragment obtained in the time interval $0\ldots t_1$. This fact leads to the appearance of a gap (discontinuity) between the last readout of the $\widetilde{x}_{out}(t)$ and the first readout that corresponds to its hypothetical continuation. Therefore, in the reconstructed solution $x_{in}^{\alpha}(t)$, there will be oscillations of values in the regions adjacent to $t=0$ and t_1 (Fig. 4.12).

Such a phenomenon of overshoot described by Henry Wilbraham in 1848 and rediscovered by J. Willard Gibbs 50 years later bears the name of the latter.

In the method under consideration, the discontinuity in periodicity is eliminated by supplementing $\widetilde{x}(t)$ with a cubic parabola $Q(t)$ (Fig. 4.13).

$Q(t)$ is a polynomial of the form:

$$Q(t)\ =q_0+q_1t+q_2t^2+q_3t^3.$$

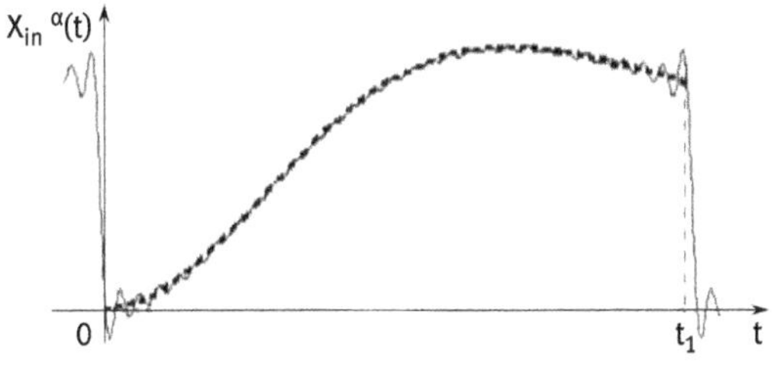

Fig. 4.12: Illustration of the Gibbs' phenomena.

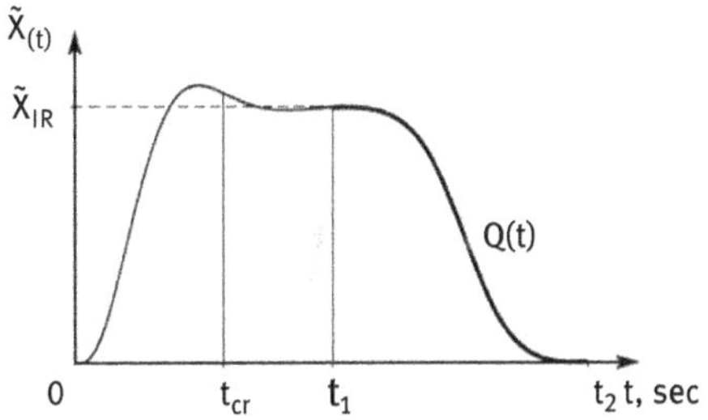

Fig. 4.13: Composition of the periodically discontinuous function.

It must satisfy the following boundary requirements:

$$Q(t_1) = \tilde{x}(t_1);$$

$$Q(t_2) = \tilde{x}(t = 0);$$

$$\frac{dQ(t_1)}{dt} = 0;$$

$$\frac{dQ(t_2)}{dt} = 0,$$

where $t_2 = 2t_1$. The equality of derivatives to zero is due to the fact that time moments t_1 and t_2 actually correspond to steady-state modes of the test object. The coefficients $q_0, \ldots, q_3$ are determined definitely from these boundary requirements.

Since the derivatives of $Q(t)$ of any order above third are equal to zero, the parameter p in the task stabilizer is limited to $p = 3$. Coefficients B_l of the stabilizer characterize the smoothness of the solution $x_{in}(t)$. Since the recovery of transient processes occurs in the frequency domain, the smoothness of the channel frequency response can be considered.

The *FRF* of the Butterworth filter is known to be as smooth as possible at passband frequencies. Hence, the values of B_l can be associated with coefficients of the Butterworth polynomial

$$B(s) = B_3 \cdot s^3 + B_2 \cdot s^2 + B_1 \cdot s + B_0 = s^3 + 2\lambda \cdot s^2 + 2\lambda^2 \cdot s + \lambda^3 \cdot$$

Here, λ is a scale factor, such that a change in its magnitude leads to a change in the time rate but does not affect the shape of the transient process. For this particular application, its value is defined as [11]:

$$\lambda = \frac{5.95}{t_{cr}},$$

where t_{cr} is a time of crossing the value of $1.05 \cdot \widetilde{X}_{IR}$ after an overshoot (Fig. 4.13) or reaching a value $0.95 \cdot \widetilde{X}_{IR}$ in overshoot absence. The time of 5.95 *sec* corresponds to t_{cr} for the dynamic system with the *TF* described by the normalized ($\lambda = 1$) Butterworth polynomial:

$$H_B(s) = \frac{1}{1 + 2s + 2s^2 + s^3}.$$

The values of the coefficients $B_l(l = 0, 3)$ obtained in this fashion are used to calculate the functional $\Psi(\alpha)$. Substituting a value of the coefficient B_1 into expression (12) allows to determined the range of values of α in which the minimum of $\Psi(\alpha)$ is sought.

4.3.3 Biases of dynamic characteristics

The effect of recovering transient processes may be illustrated by an experimental estimation of dynamic properties of the *GTE*. One of its characteristics is the graphical dependence of the corrected fuel consumption (G_{fcor}) on the corrected rotation speed (n_{cor}). This dynamic characteristic (*DC*) is represented as isobars of air pressure (P_c^*) behind the compressor and isotherms of the gas temperature (T_g^*) in front of the turbine.

The *DC* can be calculated from the results of measurements of the above parameters during the start-up of the *GTE* and following run up to the *IR* mode. Consider, as an example, the experimental estimation of the *DC* of a small size *GTE*. The rotations of its compressor shaft during start-up are represented in Fig. 4.14.

Prior to the test procedure, the dynamic calibration of the measurement channels for the abovementioned parameters was carried out. The purpose of these studies was the estimation of their *FRFs* and their subsequent approximations in the form of transfer functions.

A scheme of the dynamic calibration of a channel measuring the engine's fuel consumption is represented in Fig. 4.15.

The transducer 1 measuring G_f was the turbine flow meter. The fuel was redirected by valves 2 and 3 from the input of the engine to a spare fuel tank 4. The fuel pressure had a value corresponding to the running engine. A jump of fuel flow from

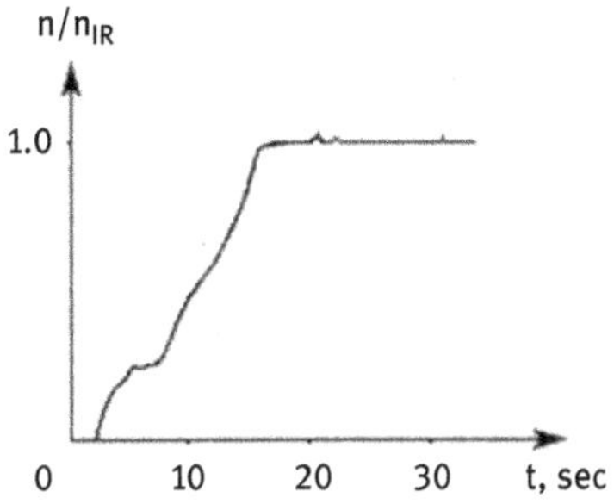

Fig. 4.14: Changing the *n* during the engine startup.

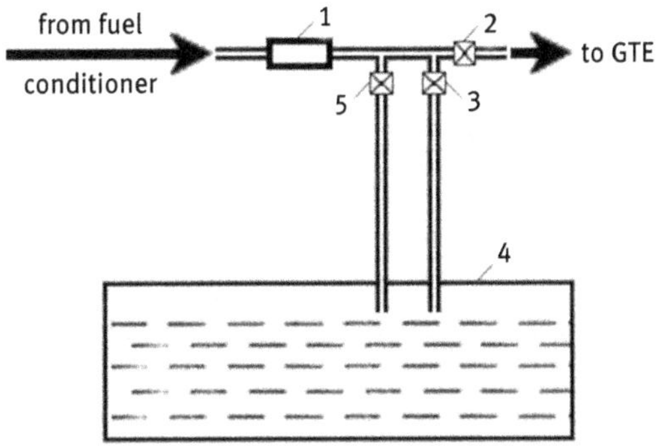

Fig. 4.15: Calibration scheme of the G_f transducer.

30% to 75% (*FS*) was created by the valve 5 ($\tau_v < 0.02\,sec$) installed in the bypass pipeline leading to the tank 4.

Processing the registered response of the measurement channel provided estimates of the frequency response. The plot (solid line 1) of the channel *FRF* is presented in Fig. 4.16.

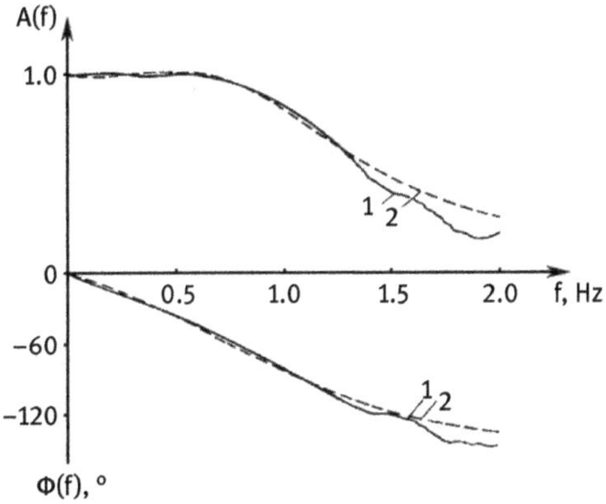

Fig. 4.16: Experimental and approximating *FRF*s.

The estimates in the frequency range of 0 . . . 2 Hz were used for approximating the transfer function. Its corresponding analytical expression was obtained in the following form:

$$H_{G_f}(s) = \frac{1.0}{1.0 + 0.21s + 0.023s^2}.$$

The approximating *FRF* is represented by the dashed line (2) in Fig. 4.16.

The channel measuring P_c^* was calibrated by supplying pressurized air to the inlet pipeline with a device shown in Fig. 4.4. Its *TF* was approximated in the following form:

$$H_{P_c^*}(s) = \frac{1.0 + 0.29s}{1.0 + 2.4s + 0.39s^2}.$$

The inertial properties of the channel measuring T_g^* are completely defined by the utilized thermocouple. The component of the *MC* following it is the *ADC*, which does not affect the inertial properties of the measurement channel. Therefore, before installing the thermocouple into the engine, it was calibrated under laboratory conditions. This approach was due to the inability to reproduce a reference temperature process in the engine.

For thermocouple calibration, a special test rig [12] was used. A sketch of its working area is shown in Fig. 4.17.

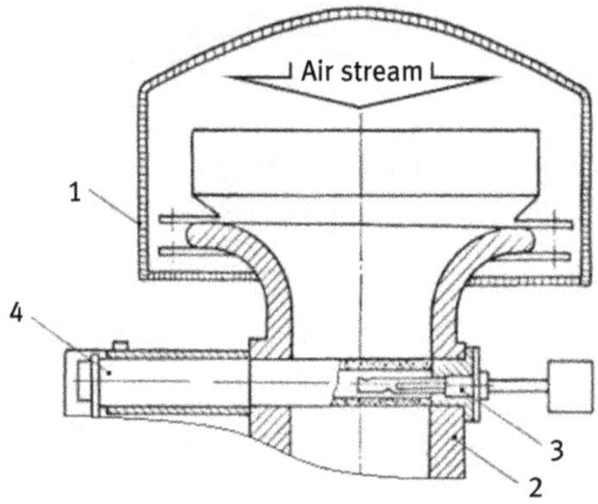

Fig. 4.17: Sketch of calibration rig.

Here 1 – air filter, 2 – vertically mounted nozzle, 3 – calibrated thermocouple, 4 – tubular electric heating element pulled in and pulled out of the nozzle by a pneumatic actuator. The out of the nozzle through the pipeline is connected with a centrifugal exhaust fan. The required velocity of the air stream is provided by changing the rotational speed of the fan. This rig provides the ability for experimental determination of transient processes of the thermocouple with the time constant $\tau \geq 0.5$sec.

The calibrated thermocouple was installed in the working section of the nozzle. The depth of its immersion in the air stream and the position of the inlets and outlets of the air stagnation chamber must correspond to operating conditions of the thermocouple in the engine.

To reproduce the transient process of the thermocouple, a heating element slides over it. Afterwards, the required velocity of the air stream and thermocouple temperature are set. When their values reach steady state, the heater is turned off and is removed from the air stream.

On this rig, the thermocouple was calibrated for a velocity of the air stream corresponding to operating conditions of the *GTE*. Its *TF* was identified in the following form:

$$H_{T_g^*}(s) = \frac{1.0 + 0.31s}{1.0 + 1.4s + 0.0018s^2}.$$

The compressor rotor speed n was measured by the alternating current tachogenerator connected to the compressor shaft through a gear with a step-up ratio (61:1). The frequency of the gear's output signal was estimated by measuring its period. Therefore, the estimates of the compressor rotor speed presented in Fig. 4.14, practically, did not have any inertial distortions.

During engine tests, the changes of the above parameters were measured when the engine was started up. Using the above *FRFs* of channels obtained from their calibrations, the measurement results were subjected to the recovery procedure described in section 4.4.2. Measured (1) and recovered (2) transient processes are presented in Fig. 4.18.

Analysis of these data unveiled considerable systematic errors (biases) of the measurement results. For instance, the underestimations of T_g^* reached 20% of the value corresponding to the *IR* mode. For estimates of G_f and P_c^*, these values were on the order of 3%.

Presented data was corrected taking into account the atmospheric conditions at the moment of tests. Source and recovered data were utilized for calculations of the engine *DC* (Fig. 4.19).

Here the dashed and dotted lines correspond to measured data, and solid lines are the recovered data. Figure 4.19 clearly demonstrates that measurement results have provided biases of estimates of the engine *DC*. Therefore, ignoring the inertial properties of measurement channels significantly misrepresents the engine characteristics, in particular, lowering engine fuel consumption.

As it is known, improving dynamic properties (decreasing startup and acceleration times) of the *GTE* is provided by increasing the excess of fuel supply as compared to the fuel supply in steady-state modes. Therefore, knowledge of actual values of fuel consumption during transient regimes is a crucial issue for creating optimal control algorithms implemented in the *ECU*. This is due to the fact that the

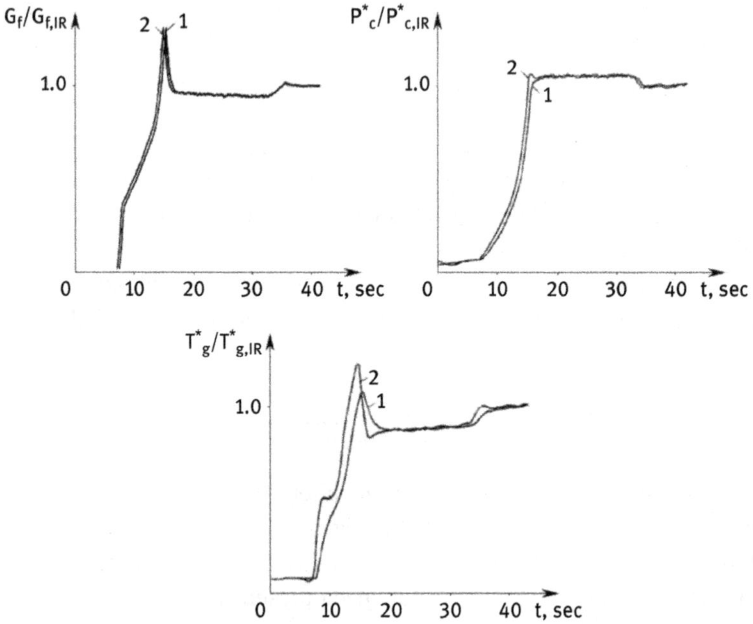

Fig. 4.18: Source and recovered data.

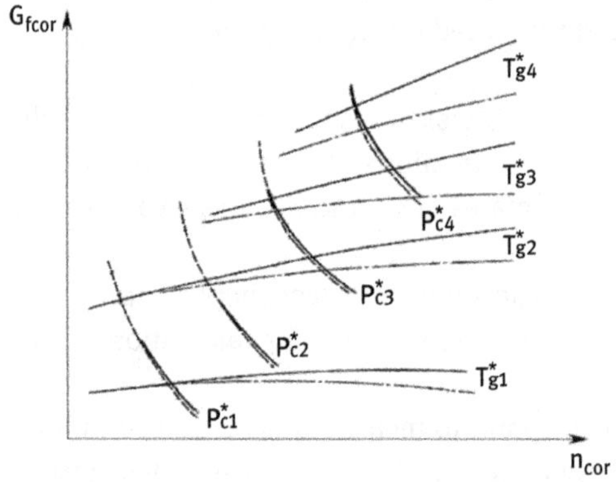

Fig. 4.19: The experimental *DC* of the *GTE*.

final stage of transient processes, as a rule, occur along the trajectory corresponding to the isotherm T_g^*.

The proposed metrological model of the *AMIS* allowed us to estimate static and dynamic characteristics of measurement channels as well as to examine their influence on test result accuracy.

Discussed approaches for evaluating instrumental errors allowed us to estimate their impact on the test results such as:

- statistical variance caused by direct measurement errors;
- systematic errors (biases) occuring as a consequence of inertial properties of measurement means.

In addition, a method to optimize the requirements for measurement errors to ensure a given accuracy of test results was discussed.

References

[1] Zacks S. Parametric Statistical Inference: Basic Theory and Modern Approaches, Pergamon, New York, 1981. 404p.

[2] Кармалита В.А., Лобанов В.Э. Точность результатов автоматизированного эксперимента. М.: Машиностроение, 1992. 208с. (Karmalita V.A., Lobanov V.É. Accuracy of Results of Automated Experiments, Machine Building, Moscow, 1992. 208p.)

[3] Jombo G., Griffiths J.D., Zhang Y., Latimer A. Towards an Automated System for Industrial Gas Turbine Acceptance Testing. 43rd Annual Conference of the IEEE Industrial Electronics Society (IECON 2017), 29 October–1 November 2017.

[4] Mazhdrakov M., Benov D., Valkanov N. The Monte Carlo Method: Engineering Applications, ACMO Academic Press, 2018. 250p.

[5] Anderson T.W. An Introduction to Multivariate Statistical Analysis, 3rd Ed., New York, Wiley, 2003. 752p.

[6] Bendat J.S., Piersol A.G. Random Data: Analysis and Measurement Procedures, 4th Ed., Wiley, 2010. 640p.

[7] van Brunt B. The Calculus of Variations, Springer, New York, 2004. 292p.

[8] Karmalita V.A., Krivonosov I.I., Lobanov V.É. Metrological support to computerized tests on aircraft gas-turbine engines under transient conditions. Measurement Technics, Volume 28, Issue 1, 1985. pp. 1–3.

[9] Isermann R., Münchhof M. Identification of Dynamic Systems: An Introduction with Applications, Springer, Berlin, 2011. 705p.

[10] Tikhonov A.N., Arsenin V.Y. Solution of Ill-posed Problems, Winston & Sons, Washington, 1977. 258p.

[11] Григорьев В.В., Журавлёва Н.В., Лукьянова Г.В., Сергеев К.А. Синтез систем автоматического управления методом модального управления. – С-Пб: СПбГУ ИТМО, 2007. 108с. (Grigoriev V.V., Zhuravleva N.V., Luk'yanova G.V., Sergeev K.A. Synthesis of Automatic Control Systems with Modal Control Method, ITMO University, St. Petersburg, 2007. 108p.)

[12] ОСТ 1 00418-81. Отраслевая система обеспечения единства измерений. Метод и средства определения динамических характеристик датчиков температур газовых потоков. (OST 1 00418-81. Industry branch system of ensuring unity of meaurements. Method and means of determining dynamic characteristics of gas flow temperature sensors.)

https://doi.org/10.1515/9783110666670-006

Index

accuracy indicator 46, 63, 65
algebraic complement 53, 55, 90
AMIS V, VI, 2, 3, 4, 33, 34, 35, 36, 37, 39, 43, 50, 67, 70, 75, 100
Analog-Digital Converter (*ADC*) 36
applied (industrial) metrology 1

Butterworth polynomial 95, 96

calculus of variations 64
calibration curve 37, 42, 43
calibration data 42, 43, 44, 45, 83, 87, 89
central limit theorem 9, 19, 46, 48
channel variation 44, 45
competitive hypotheses 30
conditional probability density 10
confidence interval 22, 43, 46, 47, 49, 51, 56, 63, 65, 82
confidence level 22
consistency 24, 26
convergence 24, 86
convex 91
convolution integral 15, 16, 90
corrected measurements 38
correlation matrix 11, 58
covariance matrix 12, 17, 18, 47, 48, 52, 53, 55, 58
Cramer-Rao inequality 26
cross-correlation coefficient 11, 52, 54, 56
cross-covariance (covariance) 10
cubic parabola 94
cutoff frequency 74, 75, 79, 81

degrees of freedom 19, 20, 29, 31
derivative 6, 13, 23, 26, 55, 72, 80, 81, 83
differential equation 72, 73, 81
Dirac's delta function 78
direct analysis 59
direct measurements V, 35, 47
dynamic system VI, 70, 71, 72, 73, 74, 78

ECU 34, 69, 88, 99
efficient estimate 24
equality of the mean values 30
error ellipse 53
error of the first kind 30
error of the second kind 30
error propagation 38, 64
experimental data VI, 1, 2, 3, 18, 21, 28, 84
experimental error 3

finite difference 80
Fisher information 26
Fisher's *Z*-transformation 56
Fourier transform 71, 73
frequency bandwidth 84
frequency domain 71, 85
frequency response function (*FRF*) 71

Gauss distribution 9, 12, 49
Gibbs' phenomena 95
GTE VI, 33, 34, 57, 67, 69, 73, 86, 88, 99, 100

Heaviside step function 74
heuristic postulate 65
histogram 31, 46, 47, 50, 51
hysteresis 41, 43

ill-posed 90, 91
impulse-response function (*IRF*) 71
indirect and aggregate measurements 35
inertial lag 67, 68, 74
inertial properties 4, 70, 78, 81, 84, 85, 89, 99, 101
inference 18, 26, 30
instrumental errors 4
instrumental noise 76, 77, 78, 80, 81, 82, 86, 87, 90
International Standards Organization (*ISO*) 37
inverse task 59, 64, 66

joint probability density 10, 12, 22

Lagrangian function 65
Laplace transform 73
least squares estimate (*LSE*) 27
least-squares method 27, 28

marginal distributions 61, 66
mathematical expectation 7, 9, 17, 19, 24, 25, 36, 42, 51
maxiimum likelihood criterion 21, 23

https://doi.org/10.1515/9783110666670-007

maxiimum likelihood estimate (*MLE*) 21
mean square error 23
measurement channel (*MC*) V
measurement data V, 2, 39
measurement error V, 27, 37, 44, 64
measurement means V, 1, 4, 5, 101
measurement results V, VI, 5, 12, 29, 34, 37, 38, 39, 41, 46, 47, 49, 50, 51, 52, 58, 59, 60, 61, 63, 64, 66, 67, 68, 69, 87, 90, 99
methodical errors 4, 48
metrological characteristics V, VI, 39, 43, 44, 47, 49
metrological structure 34
MLE invariance 23
Monte Carlo method 48, 49, 50
multiple measurements 67, 68, 69
multivariate distribution 10

normal (Gaussian) distribution 8
null hypothesis 30, 31, 32, 45, 46, 56
Nyquist frequency 75, 80, 84
Nyquist-Shannon-Kotelnikov theorem 75

Pearson's chi-squared test 31, 46
polynomial function 42, 43
power spectral density 77
PRNG 48
probability density function (*PDF*) 7
probability models 18

quantization error 36, 37

random numbers 48, 49, 51
random process 76, 77
random variable 6, 7, 8, 9, 19, 21, 31, 40, 46, 81
random vector 10, 76
recovering filter 93
recovery task 93
reference process 78, 79, 80
reference values 40, 41, 43, 44
regularizing operator 91, 92
rms deviation 40, 46, 49, 50
root-mean-square (*rms*) 8

sample & hold 68
sampling interval 75, 80, 81
stationary point 23, 65
statistical series 41, 42, 49
statistical tests VI, 39, 41, 42, 48, 49, 50
statistical uncertainty 56, 67
statistics 18, 22, 30, 56
steady-state modes 67, 70, 99

task stabilizer 91, 92, 95
test facility V, 1, 33
test of variance homogeneity 31
test results V, VI, 1, 2, 3, 4, 12, 34, 35, 38, 39, 47, 48, 52, 53, 58, 59, 64, 65, 67, 69, 101
test rig 60
throttle response 50
time constant τ 67, 68
time domain 72, 84
transfer function 72, 73, 78, 79, 85, 86, 88
transient process 94, 96, 99
transient regimes 67, 70, 99

unbiased estimate 24, 25, 26
uniform distribution 8, 14, 36, 37, 46, 48, 49, 51, 61

variance 4, 8, 9, 15, 16, 19, 22, 23, 24, 25, 26, 27, 29, 30, 31, 37, 42, 49, 53, 58, 62, 64, 69, 76, 77, 101

weight function 70, 71, 78, 80, 81, 89, 90
well-posed 90
"white" noise 77

www.ingramcontent.com/pod-product-compliance
Lightning Source LLC
Chambersburg PA
CBHW081548220326
41598CB00036B/6607